Building Chicken Coops

by Todd Brock, Dave Zook, and Rob Ludlow

D1157860

for **dummies**®

A Wiley Brand

Building Chicken Coops For Dummies®

Published by: **John Wiley & Sons, Inc.**, 111 River Street, Hoboken, NJ 07030-5774, www.wiley.com

Copyright © 2019 by John Wiley & Sons, Inc., Hoboken, New Jersey

Published simultaneously in Canada

For general information on our other products and services, please contact our Customer Care Department within the U.S. at 877-762-2974, outside the U.S. at 317-572-3993, or fax 317-572-4002. For technical support, please visit https://hub.wiley.com/community/support/dummies.

Wiley publishes in a variety of print and electronic formats and by print-on-demand. Some material included with standard print versions of this book may not be included in e-books or in print-on-demand. If this book refers to media such as a CD or DVD that is not included in the version you purchased, you may download this material at http://booksupport.wiley.com. For more information about Wiley products, visit www.wiley.com.

Library of Congress Control Number: 2018955553

ISBN: 978-1-119-54392-3; ISBN: 978-1-119-54391-6 (ebk); ISBN: 978-1-119-54388-6 (ebk)

Manufactured in the United States of America

V10006323_112118

Contents at a Glance

Table of Contents

Introduction

Which came first: the chicken or the coop? Did you get into raising your own chickens because you were lured by the idea of "free" farm-fresh eggs or swayed by your kids' desire for a few cute little fluffballs scratching about in the backyard, and only then realize that you need a place for them to actually live? Or are you just now thinking of getting into chickens and estimating housing costs as part of your research, knowing that you'd better have a proper shelter ready to go before you come home with a box full of chicks? Whichever came first for you, the desire to adopt chickens or the need to provide them with a coop, welcome.

Chicken owners are a particularly self-reliant and improvisational bunch. It's about making do and adapting. You get paid back in eggs — the equivalent of just a few bucks a month — so the vast majority of caretakers go to great lengths to keep chicken-keeping a low-cost hobby. The whole endeavor is meant to make you just a little more self-sufficient; why spend gobs of cash to do it in the first place?

Maybe that helps explain why so many chicken folks build their own coops. Sure, you can purchase a pre-built shelter for your birds (and some awfully nice ones at that), but for many, that goes against the whole reason they got into hens to begin with. Why pay for something that you can provide for yourself? And if you're clever enough and self-sufficient enough to see the benefits of raising chickens, you can indeed build your own chicken coop.

About This Book

Do some nosing around about building your own chicken coop, and you're likely to come away a little frustrated. Lots of books pontificate about how easy it is to build a working coop. Countless Web sites offer photos of shelters and first-hand accounts of the building process from the caretakers. But what almost none of them offer is instruction on *how* to do it, a soup-to-nuts guide for the marginally-handy homeowner on what a coop needs, why it needs those things, and how to actually build it yourself.

You can find all the answers you need in this easy-to-digest book. And you don't have to read it cover-to-cover. Need a quick overview of chicken coops in general? You don't have to wade through instructions on how to frame a roof to get it. Looking for a rundown of building materials to consider? It's not lumped in with step-by-step directions for constructing a chicken run. Topics are broken down into separate sections and covered in just the right amount of detail.

And in Part 3 of this book, you find something that no one else will give you: complete building plans for not one, not two, but *five* different chicken coops. They vary in size and shape, and each has its own unique features that make it special, but everything you need to construct each one of them is right there: a detailed list of what to buy, exact specs on how to cut the lumber, and precise assembly instructions.

A chicken coop seems to constantly evolve over its lifespan. It's always a work in progress. We hope this book becomes a valuable reference tool for you, even after your coop is built and your chickens have moved in. There's plenty of good advice in here on a number of issues that come up for every caretaker, and a wealth of ideas for making your "perfect" chicken coop even better.

Conventions Used in This Book

Before we get started, you should be aware of a few certain conventions — that is, standard formatting techniques — that were used in the printing of this book:

>> **Bold** text is used to highlight the keywords in bulleted lists (like we just did right there at the beginning of this sentence). We also use it to highlight the action part of numbered steps.

>> When we introduce a new term, we put it in *italics* the first time and follow it up with a simple definition. We also use italics to add emphasis.

>> All Web site addresses appear in monofont to make them stand out.

>> Sometimes, an address may break across two lines of text. If this happens, know that we have not included any extra characters (like hyphens) to indicate that break. Type the address into your browser exactly as it appears, pretending that the line break doesn't even exist.

>> We feature a lot of measurements and numerical notations. Just a reminder: Feet are sometimes expressed with a single quotation mark, as in 8'. Double quotations marks signify inches, like 16". And when discussing board lumber, the letter "x" is an abbreviation for the word "by," like when we refer to a 2x4.

>> We used 12d and 7d nails to build our coops, so that's what we reference in this book. But some areas of the country may see different sizes more commonly stocked in stores and used on job sites. If you have trouble finding 12d and 7d nails in your region, feel free to substitute 16d and 8d nails, respectively.

What You're Not to Read

Skimmers, rejoice. Not every single word in this book is absolutely necessary for you to read in order to come away with a working knowledge of how to build your own chicken coop. Sometimes we include a funny story or fascinating bit of trivia just to provide you with some interesting dinner-table conversation. Those things are put in what we call *sidebars* — gray boxes filled with text. Skip them if you want, but don't come crying to us if you make it on a game show someday and lose in the final lightning round.

And while we've tried to keep this book as accessible and easy-to-understand as possible, sometimes we had to get just a bit technical. These places are marked with a "Technical Stuff" icon, and while they may offer in-depth background information, they're not packed with make-or-break details that will ruin your coop if you don't commit them to memory.

Foolish Assumptions

Our mothers told us never to assume, but we've ignored that bit of advice. (As well as the one about waiting a half-hour between eating and swimming. That one's just dumb.) In trying to tailor this book to you, we've had to make some assumptions about who you are. Here's what we think:

>> You either already own chickens or are seriously considering owning chickens to house on your own property.

>> You already know enough about raising chickens that we don't have to spend time on the choosing, feeding, and caring of a flock. If you're new to chickens or simply want a great reference guide that deals with these issues, check out *Raising Chickens For Dummies* by Kimberly Willis and Rob Ludlow (Wiley).

>> You are either a somewhat-competent *DIYer* (do-it-yourselfer) or eager enough to learn some basic skills that the idea of constructing your own chicken coop is within your abilities. You don't have to be a highly-skilled craftsman or own a workshop full of top-end tools, but you should know which end of a hammer to hold and have a basic level of carpentry knowledge. If you need to call a handyman to hang a picture, constructing your own coop may be a bit of a stretch for you.

>> You should be familiar with schematic drawings and how to build from them (we also include step-by-step instructions to round out the instruction you get from the schematics).

>> You're not looking to become a commercial chicken farmer who needs to build an industrial-size coop. The largest coop we provide plans for can accommodate 30 birds, and the advice we offer throughout the book pertains to the backyard chicken-keeper.

How This Book Is Organized

We've tried to compartmentalize all the information in this book in a logical and organized way, to help you quickly find the piece of info you need. This allows you to dive head-first into a single particular topic and then go back to whatever you were doing. But you can also read the book straight through from front to back if you prefer.

Think of it like a multiple-course meal: If you want to sit down and start with the soup and salad, move on to the appetizer, fill up on the main entrée and accompanying side dishes, finish with a nice dessert, and top it off with a fancy cheese or cup of coffee, go right ahead. But if all you're looking for is just a quick bite of cake, we've arranged it so you can get that, too.

The book features four parts, with several chapters in each part. Each chapter is broken down into smaller, more digestible sections that are easily identified by headings in bold type.

Part 1: All Cooped Up

These chapters take a broad-view approach to what you need to know in order to start the coop-building process. Chapter 2 looks at the basics: what a chicken coop should provide, what it needs from a location, and how a few common coop styles stack up. Chapter 3 is your primer on the tools you'll need, Chapter 4 runs down some popular building materials, and Chapter 5 helps you hone your skills by teaching some solid carpentry techniques.

Part 2: Constructing a Coop

This part breaks down the building of a coop into phases. Truth be told, though, the info here doesn't apply only to chicken coops. This is solid how-to knowledge that you could put to use in building a toolshed, garden hut, greenhouse, playhouse, or potting shed.

Chapter 6 is all about prepping the site, whether that means just clearing away some yard debris or digging post holes and pouring concrete footings for an elevated structure. In Chapter 7, you tackle framing: the subfloor; the stud walls, doorways, and window openings; even roof rafters. Chapter 8 adds exterior walls, build-your-own door and window units, and a roof. We get into chicken-specific elements in Chapter 9: building a roost bar, constructing nest boxes, and adding a ramp at the door. We review runs in Chapter 10 and spark some thoughts about adding electricity to your coop in Chapter 11.

Part 3: Checking Out Coop Plans

This part features five unique coops and gives detailed instructions on how to build each one. We start with a quick, at-a-glance look at each coop's main advantages, and then provide you with a complete list of building materials, exact directions on how to cut all the lumber, and step-by-step guidance on assembling the pieces. And it's all accompanied by easy-to-read illustrations that act as helpful visual aids in the building process.

Chapter 12 offers a small coop that can be built with an absolute minimum of materials, effort, and cost. We believe it to be the simplest coop in existence. Chapter 13 features an A-frame coop that includes a run. Chapter 14's tractor coop is meant to be relocated from spot to spot in the yard. In Chapter 15, we give you a small, all-in-one coop that you can still walk into yourself. Need more room for a large flock? Turn to Chapter 16, where our biggest coop can house up to 30 birds, yet is still easy enough for a beginner to build.

Part 4: The Part of Tens

A *For Dummies* staple, the Part of Tens includes some extra takeaway info that just doesn't fit anywhere else in the book. In Chapter 17, we tell you what people who have already built a coop would do differently so that you can learn from their experiences. Chapter 18 gives you some fun wish-list items that can be added later to make your chicken-raising hobby easier, less messy, and more enjoyable.

Icons Used in This Book

Every *For Dummies* book features a collection of *icons*, special graphic symbols set off in the margins that call your attention to key pieces of information. This book uses the following icons:

REMEMBER

Remember icons point out info that should be retained for later use. If you take away anything at all from this book, it should be the information marked with this icon.

TECHNICAL STUFF

Technical Stuff icons spotlight nerdy background info or otherwise technical talk. It may or may not interest you, and skipping it should have no bearing on your coop-building adventure. (But it's great party-conversation material!)

TIP

Tips are extra scraps of info or advice. Some help you with a certain skill, tool, or technique. Some offer guidance on a specific building material or practice. Others simply provide food for thought. They're designed to save you money, time, or hassle. Maybe even all three.

WARNING

Warning icons are vitally important, because they deal with something that's potentially dangerous or harmful. Safety should be a primary concern in any building project; this icon warrants special attention and should never be skipped over.

Beyond the Book

In addition to what you're reading right now, this book comes with a free access-anywhere Cheat Sheet. To get this Cheat Sheet, go to www.dummies.com and search for "Building Chicken Coops For Dummies Cheat Sheet" by using the Search box.

Where to Go from Here

This book isn't linear — meaning you don't have to read the whole thing from start to finish. Feel free to jump around as your needs, questions, and interests dictate. But here are a few suggestions:

>> If you're just starting from scratch on your chicken career (yes, that was a deliberate pun; we do that a lot), you may want to turn the page and start with Chapter 1 for a brief overview of what a chicken coop needs to provide and what you need to consider as you decide on one for your flock.

>> If you're ready to start thinking about what materials you might want to pick up from the hardware store so you can start building, flip to Chapter 4.

>> If you'd like a basic tutorial or a refresher course on some good, solid carpentry skills that you'll use throughout the build, skip to Chapter 5.

>> If the delivery truck just dropped off a load of lumber and you're not sure where to begin, try Chapter 7 for framing assistance.

>> If you want to get right to checking out the building plans, they appear in Part 3.

1

All Cooped Up

These chapters lay the groundwork for you to construct your own chicken coop.

Chapter 1 provides a quick overview of the entire book.

In Chapter 2, we deal with logistical issues like what your coop needs to have and where it should go, and we also look at some popular coop styles.

Use Chapter 3 as a guide to the various tools you need to build your coop.

As you wade through the various building materials that are available, consult Chapter 4 for our thoughts on what's best.

Finally, Chapter 5 puts it all together by walking you through the carpentry skills you have to perform to make your coop a reality.

» Looking at tools and building materials

» Constructing your coop step-by-step

» Deciding on a coop style

Chapter **1**

Flocking to Your Own Chicken Coop

"**R**egard it as just as desirable to build a chicken house as to build a cathedral." A lovely sentiment? Yes. A bit overly dramatic? Perhaps at first glance, until you consider who said it. That quote is attributed to none other than Frank Lloyd Wright, the most famous and celebrated architect in American history. Thinking about a "chicken house" a little differently now?

You obviously take the idea of a chicken coop more seriously than most, or you wouldn't have picked up this book. While we've packed the chapters that follow with everything you need to know about how to design and construct your own coop, this chapter serves as your crash course in what you need to know to build a chicken house that even Frank Lloyd Wright would be proud of.

Understanding the Basics of Housing

A chicken coop is, at its most basic and fundamental, a shelter for your birds. It can be Spartan in its simplicity, a modest or even crude structure that serves its intended purpose but will never make the cover of *Better Coops and Gardens*. Or it can be grand and elaborate, intricately designed, and built from the finest materials, featuring all the bells and whistles imaginable.

REMEMBER

While the aesthetics may mean a great deal to you and your family as you embark on your coop-building adventure, the chickens, quite frankly, couldn't give a cluck. To your birds, a new chicken coop needs only to have a few select things going for it. These basics are explored more in-depth in Chapter 2 and throughout this book, but here's a quick list of what you need to consider before you start building a coop or settle on a specific design:

>> **Shelter:** Even wild chickens take cover when the weather turns nasty. If you're going to keep chickens in your suburban backyard, you have to give them a place where they can find shelter from rain, wind, and cold.

>> **Protection:** Humans aren't the only carnivores who enjoy a finger-lickin' good chicken dinner every now and again. A primary requirement of any coop is that it effectively offers protection from predators.

>> **Space:** We say it often in this book because it's a golden rule to always keep in mind: Your coop should provide 2 to 4 square feet of floor space for each bird you keep.

>> **Lighting:** Chickens need around 14 hours of sunlight every day. They aren't always able to get all of it outdoors. Whether it's via a window, a door, or a skylight, your coop needs to allow some light inside.

>> **Ventilation:** Chickens poop. Often. Wherever they happen to be when nature calls. The coop will get stinky. You can't prevent that, but you must exhaust that ammonia-saturated air for the health of you and your birds.

>> **Cleanliness:** Once again, chickens poop. The coop will get messy. You need to think through how you, their caretaker, will take care of that dirty job on a regular basis.

Looking at the Gear You'll Need

We'll be honest: You don't have to construct your own coop. Lots of great companies are out there who will deliver one in any size you need, ready for your flock to move into straight off the truck. Or you can easily hire a local builder, contractor, or handyman to erect one for you. The only tool you need for these options is a major credit card.

But many chicken owners love the challenge, the considerable cost savings, and the hands-on involvement of building their own coop. (We're guessing that at least one of these things appeals to you, too, or you wouldn't be reading this book.)

Building your own chicken coop may not be as easy as placing an order for a prefab unit, but it's not as difficult as you probably think, either. You don't necessarily need a garage full of professional-grade specialty gear (although a few strategically-chosen power tools can make the work easier, quicker, and more fun). We dive into tools in Chapter 3, but here's a brief checklist of the stuff you really need to have if you want to build your own chicken coop:

>> **Safety gear:** Gloves, goggles, earplugs, and a tool belt keep you in the backyard building a coop and raising chickens instead of racing to the emergency room.

>> **Garden tools:** If your coop site is currently occupied by a flower bed or a years-old pile of yard debris, you'll need to do some clearing. A rake and a shovel should suffice in

most instances. A mattock (which we cover in more detail in Chapter 3) can chop through buried tree roots.

>> **Tape measure and pencil:** Without these essential items, you're just guessing at how long a piece of lumber is or where you need to cut it.

>> **Saw:** Pick your poison — from circular saws to jigsaws, reciprocating saws to table saws, miter saws to handsaws, there are dozens of ways to cut a piece of wood. You'd better have at least one that you feel completely comfortable and fairly adept with.

>> **Tools for putting in posts:** You may need to dig a few postholes, either for anchoring timber posts that support an elevated walk-in coop or for the fence posts that define your coop's chicken run. If postholes are in your future, have a posthole digger or a power auger at the ready. (You'll probably also need a wheelbarrow and a long-handled tool like a shovel for mixing up and pouring concrete.)

>> **Hammer:** The most basic tool of them all is still the one that most coop-builders use most often. Find one you'll be able to swing all day long (but also consider a pneumatic nail gun!).

>> **Drill:** Whether you use it to drive screws or to bore small pilot holes, a powerful drill (preferably with multiple torque settings) is often the only tool that can do the job at hand.

>> **Level and square:** These tools are used in conjunction with one another as you build, to make sure that all your boards and cuts are straight.

>> **Tools for working with wire:** Wire mesh is used to enclose a chicken run or, sometimes, to cover gaps on the coop itself. A sturdy pair of tin snips will help you cut the mesh to whatever size and shape you need.

>> **Miscellaneous tools:** In addition to the basics already listed, there's a good chance you'll also find a need for things like a utility knife, a pair of sawhorses, and a screwdriver.

Choosing Coop Materials

Chicken owners, by nature, seem to be scroungers, savers, and scavengers. Chicken coops, as a result, are often constructed out of a potpourri of materials — old wooden pallets broken down into individual boards, leftover plywood from a past renovation, mismatched paint from half-empty cans in the basement, spare parts and pieces accumulated over time. These recycled and repurposed one-of-a-kind coops lend each henhouse an improvised, personal touch and are part of what makes raising backyard chickens such a fascinating hobby for so many.

But if you're constructing a coop from scratch, without the benefit of a pre-existing pile of building materials, you have some decisions to make. Chapter 4 takes a long, hard look at the different options you'll encounter at the lumberyard, building supply center, or neighborhood hardware store. In the meantime, refer to this short list of the basic materials you'll need to obtain in order to craft a coop of your own:

>> **Board lumber:** The framework of almost every coop we've ever seen is made up of board lumber. The most common cut is the 2x4, but the slightly smaller 2x3 can help you shave per-board costs and cut down on the coop's overall bulk and weight. You may need 2x6s for things like floor joists. If you're elevating your coop off the ground, 4x4s make good

corner posts. And thin boards like 1x4s or 1x3s come in handy as trim pieces for doors, windows, and various coop features.

>> **Sheet lumber:** If board lumber composes the "skeleton" of the coop, sheet lumber like plywood is often used to create the "skin." Large pieces (often 4 x 8 feet) come in thin sheets and are used for exterior cladding as well as flooring and roof sheathing.

>> **Fasteners:** To put the pieces together, you'll need either nails or screws. Both have pros and cons, and a vast array of fastener types is available to choose from. Don't overlook their importance: It would be a shame for your coop to collapse because you cheaped out on the wrong kind of nails.

>> **Flooring materials:** Almost all coop owners cover the floor of their shelter with some sort of loose bedding, like pine shavings. But underneath that bedding, many coop floors feature a smooth layer of linoleum (or a similar product) to make cleanup even easier. Some coop setups may utilize a concrete or dirt floor.

>> **Materials for walls:** The coop's solid exterior walls are most often made from sheets of thick plywood, either smooth-surfaced or with vertical grooves to create a paneled look. If you'd like to use a siding product similar to what you'd use on a house, see Chapter 8.

>> **Roofing materials:** Shingles are the classic choice for a roof, but many coop-builders use large corrugated panels of metal, fiberglass, or PVC to encourage rainwater to shed away from the coop structure.

>> **Wire mesh:** This material is so closely associated with chicken coops that "chicken wire" has become a catchall term that some use to refer to any type of flexible, metal-wire mesh. It's used primarily to enclose runs or to provide an open-air screen for the windows or doors of a shelter.

>> **Posts:** Whether they're supporting the entire structure of an elevated coop or used in a fencing application on a chicken run, posts need to be beefy enough to support the load. The most popular builder's choice is 4x4 lumber.

Getting Up to Speed on Carpentry

You don't have to be Bob Vila (or Ty Pennington, for you younger readers) to construct a quality coop that your chickens will love and you'll love to show off. But you do need to have a handle on some basic carpentry skills that are instrumental in any building project.

TIP

If you're a do-it-yourself (DIY) rookie or doubt your carpentry skills, take advantage of the years of experience and volumes of knowledge of the employees working the aisles of your local hardware store or home center. They're usually more than happy to walk you through a specific skill or teach you how to use a certain tool. Some of the larger building supply warehouse stores even hold free clinics on all kinds of how-to topics and let you try out a tool or technique in a safe, supervised environment.

REMEMBER

We've devoted Chapter 5 of this book to the skills you'll want to master before kicking off your coop-building project. Take a look at this list to see what you may need to brush up on before the sawdust starts flying. You should know how to

- **» Accurately read your tape measure:** Reading the big, fat numbers is easy. But can you differentiate at a glance between 7⅝ inches and 7¹¹⁄₁₆ inches? The difference is only the width of this capital F, but it could inspire a few choice words that start with that same letter when two pieces don't fit together because you guessed wrong.

- **» Precisely mark materials:** From making simple slash marks with a pencil to snapping chalk lines, how you mark a piece for cutting usually determines how accurate the cut is. Mark with a "V" for accuracy, and use an "X" to identify scrap ends.

WARNING

- **» Safely use a saw to cut lumber:** Power saws can make short work of a 2x4 or sheet of plywood. But they can also make short work of your index finger or thumb if you're not careful. Check for obstacles in the saw blade's path before starting a cut. Understand how to hold a saw, where to look at your workpiece for the best view, and how to stand during the cut to maintain good balance. Always properly support the piece you're cutting.

- **» Properly use a hammer:** Banging a nail flat into a piece of wood is one thing. Gripping the hammer low on the handle and swinging from the elbow instead of the wrist can make it an even easier thing. Toe-nailing a nail into a tight corner or using the claw end to pull out a mistake takes your nailing know-how to a whole different level.

- **» Read a level:** A simple carpenter's level shows whether the piece you're installing is perfectly horizontal *(level)* or precisely vertical *(plumb),* but only if you can interpret what the bubble in the vial is telling you.

- **» Use a square:** The speed square is a versatile tool that can lay out straight pencil lines, establish perfect 90-degree angles, and act as a straightedge or cutting guide for your saw. You can also use the etched markings on the angled side to mark an angle — anywhere from 1 to 89 degrees — with ease, an invaluable skill when laying out and cutting roof rafters.

- **» Use a drill:** Whether you're driving wood screws during framing, boring pilot holes in a stubborn piece of lumber, or attaching hardware at project's end, using a drill is usually as easy as squeezing the trigger. But you should be familiar with your drill's particular torque settings and other features before embarking on a big building project.

Constructing a Coop: The Nuts and Bolts

The building process, for a chicken coop or anything else, is rarely a quick one. Nor should it be. "Haste makes waste," as they say, and if you try to hurry your way through coop construction, it'll almost certainly show in the end. Be realistic about how long the build will take . . . and then add some additional time on the back end for good measure. If you estimate it'll take you two weekends, plan on a third just in case.

Part 2 of this book breaks down the build into phases. Not every phase applies to every coop design, so you may be able to skip a phase here or there. But generally speaking, the following sections give you a step-by-step rundown of how to construct your chicken coop.

REMEMBER

Coordinating your construction efforts with your chickens' readiness can be tricky, but considering this factor is critically important. The day you bring home a box of fully-grown adult chickens is not the day to start thinking about what kind of coop you want to build. The ideal scenario is to have your coop built and finished just as your chicks are ready to move in.

It doesn't always work that way, of course, so be sure to make some temporary housing arrangements for your flock before you begin building their permanent housing.

Readying the site

It starts with picking the perfect location, something that Chapter 2 deals with at length. But the ideal spot may be on a rocky portion of uneven ground, underneath a massive tree with an exposed root system and low-hanging branches, or even on a sloping hillside.

If you're lucky, you'll have very little site prep to do — maybe just some light debris cleanup, relocating a few plants, or a bit of minor regrading of soil. If you have a nice patch of flat ground, you may not have to do any site prep at all.

But you may need to consult Chapter 6 for a look at how to level the ground using stakes and a string level. On more serious slopes or for a coop that will be elevated on permanent posts, you'll have to do some heavy lifting. You may need to dig post holes and pour concrete to create *footings* — concrete pillars buried in the earth that support your structure's timber legs.

TIP

Of course, some chicken coops don't occupy just one spot in the yard. Many caretakers move their coop from place to place on the property to let the birds work different patches of ground with their pecking and scratching, or just for a change of scenery (yours as well as theirs). While some "tractor" coops are built on heavy skids or with wheels, many smaller coops can also be moved around pretty easily if they're solidly built to begin with. Of the five coops we provide plans for in Part 3, three of them are designed to be portable. Chapter 2 introduces tractor coops and other styles.

Framing

The framing may not be the part of your coop that you'll see every day as you look out your kitchen window, but it's quite literally the backbone of your chickens' housing. The frame is the skeleton that everything else is built on, so if it's not solid and sound, you may be in for a difficult build and long-term problems with the coop's stability.

The framing is made up of three basic parts: a floor, walls, and a roof. On all three parts, you typically begin by constructing a network of framing members, most often 2x4s:

>> **The floor:** Floor joists provide support for a solid decking material that becomes the floor.

>> **The walls:** Vertically-arranged studs provide rigidity for exterior walls; doorways, window openings, and access hatches all have their own stud framing that gives them the strength to stand up to daily use.

>> **The roof:** Rafters give solid backing to sheets of roofing material.

TIP

Framing is the most basic of carpentry tasks and a great place for the building beginner to practice sawing, hammering, and leveling skills. While being accurate with measurements and cuts is important, most framing is covered up by something else, so absolute perfection is seldom required. (But Chapter 7 has everything you need to know about how to get it as perfect as you possibly can.)

Putting up walls, a roof, and more

After you've completed the framing of the coop, you'll be ready to enclose the shelter with exterior surfaces. Chapter 8 details the ins and outs of adding walls, doors, windows, a roof, and ventilation to your coop. Here are some basics:

>> **Walls:** Plywood is the wall material of choice for most coop-builders, but different kinds of siding can be used instead (although some have their own specific framing requirements).

>> **Doors:** Nearly every coop has to have some kind of access door — for humans, chickens, or both. The majority of coop-builders choose to build their own door from scratch, using the same general framing and cladding skills that they've demonstrated up to this point in the build. This allows the DIYer to custom-make a door that perfectly fits not only the opening in the coop, but the way it will be used day in and day out.

>> **Windows:** When it comes to windows, some people like to keep it super-simple by fastening wire mesh over an opening for a permanent screen effect. Many folks build mini-doors or hatches that can be propped open or latched closed. Others use actual window units like you have in your own house; many of these were once working house windows, now recycled for a second life in a coop.

>> **A roof:** First and foremost, a roof must keep the interior of the shelter dry. Plywood sheathing and asphalt shingles do the job for most coops, although corrugated panels offer some nice benefits that appeal to many chicken-owners.

>> **Vents:** Ventilation is often built into the roof structure by way of ridge vents or the fancier cupola, but many coops add simple vents into the shelter walls to help stale, stinky air escape the confines of the coop.

Adding special touches

At this point in the building process, you have a big box. It takes a few extra creature comforts to turn that box into a chicken coop. The following easy-to-build pieces are explored more thoroughly in Chapter 9:

>> **Roosts:** A *roost* is, at first glance, a simple bar or pole inside the shelter that replicates the tree branch that your birds instinctively want to perch on at night. Provide 12 inches of roost per bird, and place the roost bar as high off the ground as your coop design allows. Steps (or other roosts to be used as steps) can often help the hens make their way up to their perch at night.

REMEMBER

The area directly below the roost gets the most chicken poop, so pay extra-special attention to where you place your roost. Open-top nest boxes and food or water containers should never be placed in the "drop zone." In fact, savvy coop owners design the coop specifically with the logistics of future cleanups in mind.

>> **Nest boxes:** Birds being kept for eggs should be given nest boxes. Hens will share a nest box, so plan on having one for every two to three birds. Each box should be a minimum of 12 x 12 inches, or larger if possible. Nest boxes can be completely contained within the shelter, but many caretakers build a bank of nest boxes that stick out of an exterior coop wall, with a lid that can be lifted from outside the shelter to access the eggs.

>> **Ramps:** Some coops may require a ramp that offers the flock a way to get down out of the coop and into the run or yard. These simple ramps often feature rungs that give the birds' feet something to grip as they make their way up and down.

Building a run

Of our five coop designs in Part 3, three incorporate a run into the structure itself. But a stand-alone shelter needs an attached run, an enclosed outdoor area where the chickens will spend their days hunting and pecking. Factor in 3 to 6 square feet of run per bird, although more is always better.

First-time coop-builders should think of the fencing material that makes up the run not only as something that keeps chickens in, but also that which keeps hungry predators *out*. Light-weight netting or flimsy chicken wire is rarely enough; plan on stretching heavy-duty, welded wire with small openings between posts that are solidly anchored and securely fastened.

REMEMBER

Many caretakers also rig up fencing material over the top of the run to thwart airborne attacks. See Chapter 10 for much more about building a run.

Hooking up electricity

Running an electrical line to a chicken coop is an expense that many caretakers are reluctant to consider. Chicken owners who already have power in their poultry pen will tell you it's worth every penny. After reading Chapter 11, you might think so, too. From simple task lighting to space heaters to turbine fans that exhaust stale air, electricity can spark all kinds of ways to make your coop cleaner, better, and easier to maintain.

Messing around with wiring, though, is a potentially dangerous endeavor for many weekend warriors. If you like the idea of having juice in your coop, think about hiring a professional electrician to take care of it.

WARNING

Checking Out a Few Coop Designs

Part 3 is what really sets this book apart. That's where we provide full building plans for five different chicken coops; at least one should fit the needs of any urban chicken-keeper. Chapters 12 through 16 contain full materials lists, cut lists, and assembly instructions — all with detailed illustrations to guide you through each cut and every connection.

In a nutshell, here are the five coops you'll be able to build by the time you reach the end of this book:

>> **The Minimal Coop:** Our smallest coop, in Chapter 12, is for the do-it-yourselfer who doesn't want to do all that much. It uses basic pieces of lumber and requires only simple straight-line cuts — and as few of both as humanly possible. This no-frills "starter" coop houses four to five birds but doesn't include a run.

>> **The Alpine A-Frame:** Another small coop, for two to four birds, the plan in Chapter 13 requires even less material to build the shelter, thanks to an A-frame design. A hinged roof panel allows easy caretaker access. A 24-square-foot run is attached to this self-contained coop, and attractive exterior siding makes it a nice addition to the landscape.

>> **The Urban Tractor:** Specifically designed to be relocated around your property via a heavy tow chain, this coop (featured in Chapter 14) incorporates a shelter that accommodates two or three birds and a 16-square-foot run in a portable unit. Part of the run extends underneath the elevated shelter to provide your hens a shady spot to chill out on hot summer days.

>> **The All-in-One:** Combining the best features of all coop styles, this coop (shown in Chapter 15) is small enough to be portable, yet tall enough that the caretaker can step into the attached run and stand upright. The shelter can accommodate four to six chickens, yet occupies a smaller footprint than the Alpine A-Frame coop.

>> **The Walk-In:** Our largest coop (shown in Chapter 16) is for chicken owners who want to start out with a big flock or just increase their existing bird count. The 8-x-8-foot shelter can house up to 30 birds at once (or fewer birds with some storage space leftover!). Building it, though, doesn't require any special skills that aren't already explained in this book.

Chapter **2**

Beginning with Housing Basics

I t's the age-old question: which came first, the chicken or the coop? (Or something like that.) But seriously, do you really need a chicken coop for your backyard flock? Humans were keeping chickens long before they started building little houses for them. So why is dedicated housing even necessary?

Technically speaking, a chicken owner could let a bunch of hens fend for themselves with no coop whatsoever. It's *possible*. Unfortunately, it's not *practical*. For almost every one of us, no matter where we live, having some sort of housing is a prerequisite to keeping chickens.

"Some sort of housing" is open to interpretation, and many a chicken-owner has improvised his way to a serviceable shelter. A toolshed, an outgrown playhouse, or even a repurposed doghouse can easily be retrofitted with a roost and a nest box and turned into a darn fine chicken coop.

But for a large number of chicken-owners (and maybe most), the desire to keep a backyard flock comes hand-in-wing with an urge to construct a coop. Chalk it up to the pioneering, self-sufficient streak that runs a mile wide in most folks who decide to get into chickens as a hobby, but for many, getting out there with a bunch of tools and building a personalized, one-of-a kind shelter is one of the best parts of the whole chicken-raising endeavor.

Coops come in all different shapes, sizes, layouts, and configurations. The perfect coop for you depends on the size of your flock, the space limitations of your property, the rules (either official or implied) of your neighborhood, the amount of work you're willing to do, what predators frequent your area, your climate, and your reason for keeping chickens in the first place. Although we're confident that one of the five coops we feature in Part 3 will work for you, there's no such thing as a "one-size-fits-all" coop.

All chicken coops, whether improvised, assembled from a prefab kit, or constructed from scratch, must meet some elemental requirements. And there are several factors that you must consider before making any decision on a coop in your own backyard. This chapter runs down those requirements and considerations, and points out how a few common coop styles stack up.

Providing Basic Benefits with Your Coop

REMEMBER

Whether the coop you have pictured in your mind is a basic box cobbled together from some scrap wood or an enormous architectural masterpiece with its own zip code, a coop must provide the following:

>> **Shelter from the weather:** A chicken coop, at its most basic, is a shelter from the elements. It's a place for your flock to stay dry in wet weather, get warm in cold weather, and grab some shade in hot weather. Just like people, chickens don't particularly enjoy being exposed to rain, wind, snow, or extreme temperatures. Their coop is where they go to be protected from whatever Mother Nature dishes out.

>> **Protection from predators:** The coop is your chickens' refuge from not only weather, but also predators. Even the most serene and idyllic setting is chock-full of animals who would kill (literally) for a claw-lickin'-good chicken dinner. Raccoons and opossums are prevalent chicken-hunters. Foxes and coyotes are more common than you might think. Skunks, weasels, and minks aren't unheard of. Hawks and owls can bring terror from the sky. In some regions, snakes, cougars, bobcats, and even alligators are a big problem.

But the biggest nuisance to chicken-kind? Fido, Rover, and Spot. Dogs love to chase chickens, and they typically end the chase with a kill. And we're not just talking about that mean stray dog that occasionally wanders through your cul-de-sac. Even domestic pet dogs, no matter how well-behaved or mild-mannered, will often make short work of a backyard flock if given the chance. A coop not only keeps your chickens *in,* it keeps these predators *out.*

>> **Space:** Remember that tiny one-bedroom efficiency apartment you suffered in through senior year of college? You wouldn't wish that on anybody, right? Well, don't cram more chickens in your coop than it's meant to hold, either. A good rule of thumb is 2 to 4 square feet of floor space inside the shelter per adult chicken; this amount of space allows chickens to strut about normally, flap their wings, and socialize properly with their coopmates. (See the later section "Understanding that yes, size matters" for more about providing plenty of room for your chickens.)

>> **Lighting:** Chickens need light to function, and they don't do well in darkness (even at night!). While some coop-builders like to incorporate electrical lighting for their own needs and convenience (as described in Chapter 11), most large coops simply have a window or two to let the sun shine in. In a smaller coop or where a window isn't an option, it's important to provide natural light via an outdoor run (see Chapter 10 for more details on runs).

>> **Ventilation:** Fresh air is a necessity. It doesn't take very long for the inside of a coop to get downright funky after you introduce a few hens and their potent droppings. The ammonia-saturated air that results is bad for your chickens and can lead to respiratory illness, not to

mention mold and mildew, and just a general unpleasantness about your coop. You can either design some simple vents into your coop (as explained in Chapter 8) or go with a plug-in fan alternative, but how you'll exhaust stale, ammonia-saturated air from the shelter should be something you think about sooner rather than later.

>> **Cleanliness:** Chickens like things tidy. So unless you think you can potty-train your flock and teach them how to work a broom, you'll have to get out there yourself and do some cleanup from time to time. That may mean nothing more than an occasional scoop-out with a shovel or spray-down with a garden hose, but give some thought now, during the planning stages, to how you'll clean your coop in the long run. (Check out Chapter 4 for more on flooring materials you may want to consider.)

>> **Temperature control:** Chickens generally do best when the mercury's between 40 and 85 degrees Fahrenheit. Many coop-builders, then, don't need to worry too much about controlling the climate inside the shelter. But if you live somewhere that sees prolonged heat waves in summer or particularly harsh winters, you may want to start thinking about how you'll help keep your birds comfortable. (Chapter 11 has plenty of cool ideas and advice you'll surely warm up to.)

Analyzing the Anatomy of a Coop

Even though coops come in a myriad of shapes and sizes, with special amenities and added quirks as varied as the chicken–owners who build them, there are a few basic features that every chicken shelter should have. Think of the following items as furniture for your coop, the physical things that take an empty structure and turn it into a chicken's home:

>> **Roosts:** A *roost* is where your chicken will sleep. True free-range chickens in the wild find a tree branch to perch on each night to sleep. This offers them protection from non-climbing predators, and a branch under the canopy of a tree can provide cover from hungry owls and nasty weather. In your coop, a horizontally-oriented pole or board that's elevated off the floor serves the same function, as seen in Figure 2-1.

FIGURE 2-1:
A simple roost is a chicken's master bedroom.

Roosts are explored more in-depth in Chapter 9, but for now, aim for giving each bird at least 12 inches of roost. Your roost (or roosts) should be as high as possible inside the coop while still allowing your chickens to sit fully upright on it. (But because chickens aren't big flyers, they may need some help reaching the roost, especially as they get older. A ramp or series of horizontal "steps" to the roost bar can be used as needed.) Also, realize that the area directly under the roost will collect the greatest concentration of droppings, so plan accordingly.

» **Nest boxes:** If you're keeping chickens for their eggs, nest boxes are a necessity (even though many owners of particularly-finicky chickens will tell you that their nest boxes go unused more often than not). A *nest box* (see Figure 2-2) is exactly that: a safe, comfortable, and secluded spot where your hens can lay their eggs.

FIGURE 2-2: Nest boxes are critical if you're encouraging egg production from your flock.

Chapter 9 offers step-by-step directions on building your own nest boxes, but keep a few general rules in mind as you plan:

● Hens will share nest boxes, so having one nest box for every two (or even three) birds is sufficient.

● Each box should be a minimum of 12 inches square, or big enough so a chicken can comfortably walk in, turn around in a circle, and then stand fully upright when she's done laying.

● A nest box should be in a dimly-lit and warm area of the coop, no more than 3 or 4 feet off the floor.

● If you have more than one box, bunch them together to allow the birds to socialize with one another as they do their business.

» **Runs:** A *run* is the yard for your chicken coop, giving your flock access to fresh air and sunshine. It's an open-air space, usually wrapped in some sort of fencing material (as discussed in Chapter 4) and may even be fully covered by a roof. A run should allow each bird to have 3 to 6 square feet, but more is better. For most backyard chicken owners, a coop that incorporates the run into one structure is generally easiest to build and maintain, but folks with enough land may prefer to have a large fenced run that encompasses the chickens' shelter, or even allow the birds to roam free-range with no fencing at all. There's plenty more on constructing chicken runs in Chapter 10.

WARNING

Look out below! Chickens hanging out in their run may be susceptible to predators that can burrow underneath the frame. Consult Chapter 10 for an easy way to safeguard against subterranean attacks in a stationary run; it requires just a little bit of shovel work on your part during the build.

>> **Ramps:** Chickens are birds, but that doesn't mean they soar through the air like eagles. In fact, chickens don't do much flying at all. Many coops are built with a sizable (for a chicken) drop from the shelter down to the ground. Your chickens will be much happier if they don't have to flutter and flop their way in and out of the coop on a regular basis. Look at constructing a simple ramp that lets your flock strut up and down in style. It usually requires no more than a few pieces of scrap wood, but it'll make all the difference in the world to your birds. (Look for more ramp-related revelations in Chapter 9.)

Making Your Coop Convenient for You

It's one thing to go on at length about what the chickens want and need out of a coop. But factoring yourself into the equation is just as important. Unfortunately, many newbie owners go strictly by the book when it comes to their coop's location, size, style, and features . . . and later find themselves quickly falling out of love with chicken-raising as a hobby because the coop is too far away from the house, too difficult to clean, too small, too big, or too whatever. Often, just thinking through a particular aspect of a design and your particular circumstances will point you in the right direction.

Most chicken-owners will tell you that keeping a flock and maintaining a coop is always a work in progress. There will almost always be a better way to do something or a new trick you can employ to make some part of the process easier, better, quicker, or more rewarding. Feel free to tinker with your coop, during the design phase and even long after it's built and inhabited.

TIP

Having a tough time chasing down hens because your roosts are in the way? Attaching one end of each roost to the coop wall with a hinge will fix that. Is cleanup taking too much of your time? Try adding some litter pans under the roosts or constructing a giant slide-out tray that can be removed for cleaning. And chicken owners everywhere owe a debt of gratitude to the first guy who thought to build his nest boxes into the wall of the coop, with lids that could be opened from *outside* the coop for easy egg-collecting!

The point is, those kinds of innovations weren't in the very first chicken coop; they were developed over time as flock-keepers and coop-builders saw ways to make improvements. We point out coop-specific tips here and there that can make your job as a caretaker easier (Chapters 17 and 18 contain loads of ideas), but understand that it's also a learn-as-you-go experience, with lots of trial-and-error and on-the-fly improvisation.

Selecting Your Coop's Location and Size

Location, location, location. It's true in real estate, and it's equally true when it comes to chicken coops. Location dictates everything.

Before you flip ahead to the next section and fall head over heels for a particular coop style, you need to begin by determining where on your property a coop can and should go. This section tackles a few of the considerations you should weigh carefully in deciding which corner of the yard will become home to your hens.

Before you begin: Considering zoning and covenant concerns

Before you ask, "*Where* should I have chickens?" you really need to first ask, "*May* I have chickens?" Despite all of the advantages and benefits that come with raising a backyard flock, some places simply won't allow it.

It could be that your property's zoning (as classified by local governments) doesn't allow chickens to be kept on the premises. Some neighborhood covenants or homeowners' associations prohibit backyard chicken-raising. In some cases, the law may say nothing about keeping chickens, but will come down on you like a hammer when you go to construct a coop without proper permits or inspections.

REMEMBER

Just because a neighbor in your area has chickens and/or a mighty fine backyard coop does *not* automatically mean that everything is cool for you to do the same. His property's zoning may be different from yours. He may have started raising chickens before some change in local laws or the advent of the neighborhood's covenants. He may have gone through proper channels to obtain a zoning variance. Or it may just be that he's in violation of the law himself and keeping chickens illegally.

Do your homework and know what's acceptable before investing your time, effort, and money in a backyard flock or a built-from-scratch chicken coop. Be sure to check with local government officials about zoning restrictions and whether building permits are needed for constructing a coop on your property. This may mean a trip to your city or township hall, the community planning board, the county clerk's office, or even animal control! And even if *they* all give you the green light, make sure your neighborhood or subdivision doesn't have a rule against a backyard chicken coop.

Looking at proximity to houses

Assuming that it's perfectly legal for you to raise chickens in your backyard and house them in a coop, you'll want to select a patch of your property that's close, but not too close, to your own house — or your neighbors'.

Regarding your house

One of the most common mistakes first-time coop-builders make is putting the coop too far away from the house. In an effort to provide the chickens with sufficient room and adequate cover from predators, while making the structure blend in with the landscape so it's not an eyesore for neighbors, they tuck that coop in the very back corner of the yard. Well, as the old saying goes, "Out of sight, out of mind." A faraway coop location just makes it more difficult and time-consuming to go out and check on the flock. (You'll need to do this at least twice a day every day!) If it's raining or bitterly cold outside, that long trek to the coop will only encourage you to come up with reasons to *not* raise chickens.

If the size of your property allows, aim for someplace that's within sight of and an easy walk from the house. This allows you to keep a lookout for predators (and escapees) while making maintenance less of a chore.

WARNING

But don't go too far the other way, either. You probably don't want to put your brand-new chicken coop right up next to the house. Sure, "closer" means "more convenient." But it also means "smellier" and "noisier." And those pretty flower beds all along your home's foundation? Chickens will make short work of those in no time, so keep your flock well away from any landscaping you want to keep.

Considering your neighbors

Many neighbors of chicken-owners are supportive and encouraging (especially if it means a few fresh eggs every now and again!). Many others, though, aren't exactly crazy about having Green Acres right next door.

Either way, it's probably not a good idea to locate your coop too close to your neighbors' houses. Why?

>> First, there's the odor issue. Enough said.

>> Then, there's their view to consider. More than likely, they've put a good deal of thought and planning into their perfect backyard just like you have. Is it really fair to make them stare at the backside of your chicken coop every time they look out their bedroom window? Or to subject them to an eyeful of your chicken run during their next cookout? (Come to think of it, that's a little cruel to your flock, too, should barbecued chicken happen to be on the menu.)

REMEMBER

Taking your neighbors into consideration when it comes to selecting your coop location may be the smartest move you make during this whole adventure. Creating animosity amongst neighbors makes life unpleasant for everyone, and having your backyard hobby be the root cause is a surefire way for your chicken-keeping endeavor to end badly. Talk over your plans with those on either side of (and maybe behind) your home, and offer them a chance to throw in their two cents. If they're cool with it, you look like a hero for being considerate. If their feathers get ruffled, you can factor that in before you build — and even gather proper paperwork that shows you're within your rights if they try to play hardball.

Utilizing utilities

While scouting out the location of nearby houses in relation to your planned coop site (as explained in the preceding section), take a quick look around for two other things that will make your life as a chicken-keeper much easier: plumbing and electricity.

Plumbing

Ah, running water. The kind of thing you take for granted until it isn't there anymore. You'll obviously need to provide your flock with fresh drinking water and change it out regularly if you're not using an automatic watering system. In addition, you may find that you like having an ample supply of water handy for coop-cleaning purposes.

It may seem like a no-brainer to locate your coop within easy reach of a garden hose. But you'd be surprised how many coop-builders don't think about it until they're lugging pails of water back and forth across the yard.

TIP

No hose spigot conveniently located near your coop? Put one there! A faucet extender (see Figure 2-3) attaches to the spigot at your house with a short hose that's typically 5 to 10 feet long. (It can be even farther away if you use a full-length garden hose as the connecting piece.) A steel stake driven into the ground holds a shut-off valve at a handy height. Turn on the water at the house; then plant this great gadget just a few steps away from the coop for running water at your fingertips — all without ever calling a plumber!

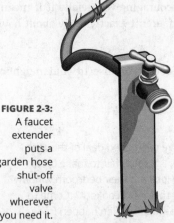

FIGURE 2-3:
A faucet extender puts a garden hose shut-off valve wherever you need it.

Electricity

Your birds won't be staying up late to read the latest chick-lit page-turner, they won't be watching *Pecking with the Stars* on their plasma flat-screen, and they won't be surfing their way over to backyardchickens.com for lively online discussion and all the latest tips. So unlike our own shelters, a chicken coop doesn't really need dedicated electrical power.

But some owners like to provide permanent, hard-wired lighting inside a coop, whether as supplemental lighting for the hens or just as a work light for themselves. (Check out Chapter 11 for more on whether juice is justified for your particular coop setup.)

REMEMBER

If you plan to include actual electrical wiring in your coop, give some thought to how far the electrician doing the work will have to run the wire, and know that a longer run translates to a more expensive service call.

Even if you don't plan to have any electricity inside your coop, there may be times when a nearby outlet would be handy. Maybe you'll want to plug in a small milk-house heater to keep your flock from freezing on a bitterly cold winter night. Perhaps you'll need to do some on-site work one day and power up a reciprocating saw to cut a new window in your coop. Or it could be that you'll just want to plug in a radio and rock out while you tackle some coop clean-out. Putting your coop within an extension cord's reach of an outdoor receptacle makes each of these things possible and easy.

Digging into drainage issues

Any time a professional landscaper starts to plan a new project, the first thing he or she will examine is the site's drainage. Because your coop will be outdoors and technically part of the landscape, the way water moves and drains on the site is something you should take time to consider, too.

Chickens don't like standing in muddy water for hours on end any more than you would, so locating your coop or run in a shallow depression that collects water after a rain won't go over well with your flock. Siting your shelter at the bottom of a slope can be a recipe for disaster, too, because heavy rains can wash unwanted debris right into your chickens' home sweet home.

TIP

If a slope or low spot in the yard is your only choice, you may want to erect your coop on heavy-duty support posts that are anchored in the earth with concrete pilings. Advice on how to do this can be found in Chapter 6.

Just as you'll want to mull over what could drain *into* your coop, pay some attention to how things will drain *out of* your coop as well. When you go to hose the place out, all that manure will be washed somewhere. Make sure that "somewhere" isn't a pond, lake, stream, or other water source where pollution is an issue. And if your home draws its water from a well, the chicken housing should be (at the very least) 50 feet away to prevent contamination of the well water.

TIP

Many smaller coop styles are designed to be portable and move from spot to spot within your yard. If you're having trouble finding one location on your property that fits all of the preceding requirements, one of these tractor coops (see the later section "Chicken tractors") may be just the ticket. You can move the coop close to the house as needed, back it into a corner when the next-door neighbors have a garden party, haul it to the hose spigot on clean-out day, and relocate it to higher ground before that big thunderstorm comes.

Understanding that yes, size matters

The size of the coop you build is often decided by the amount of space you have to work with (after you've figured out a good spot according to the considerations we discuss in the previous sections), and that, in turn, dictates how many chickens you can house. For example, if you live on a tiny, in-town lot, a massive chicken barn probably isn't going to happen. You're more likely looking at a small, self-contained coop and just a handful of hens.

REMEMBER

Figure on a minimum of 2 to 4 square feet of floor space in the coop itself and 3 to 6 square feet of outdoor run area per bird. For four hens, you'll need a coop that's at least 8 to 16 square feet and a run that's 12 to 24 square feet. So any patch of land that's between 20 and 40 square feet will suffice. If the only available spot in your yard is 16 square feet, you'll need to down-size the coop (and possibly the flock) accordingly. Or get creative with the design. Use the 16 square feet of ground for the run and then build a 16-square-foot coop on stilts, with the housing built directly over the run! (Your hens will love the sub-shack shade!)

In addition to calculating the square footage of the coop's floor space, you also need to decide how tall you want your coop to be. Many coops are just tall enough for the chickens. These may require you to do some crawling on your hands and knees to collect chickens or perform maintenance. Some even incorporate a hinged roof in the design for human access and maintenance.

But many chicken-owners build walk-in coops for the ultimate in convenience. Just like a toolshed or potting shed, they feature a human-sized door and a floor that can accommodate human weight. A walk-in, shed-type coop certainly makes cleaning and maintenance easier, because there's simply more room for you to maneuver about inside. This type of coop often complements the landscape better than a small A-frame coop. And many chicken-owners find the extra space of a walk-in handy for storing some of their chicken-related supplies. But building one does require more materials and demands extra care when selecting materials and constructing the shelter. (See Chapter 4 for more details on building a coop you can literally get into.)

Choosing a Coop to Fit Your Needs (and Skills)

Once you've considered the nonnegotiable requirements that a coop must meet, thought about a few of the extras that your flock will appreciate, mulled over how you want to care for and maintain a shelter, consulted with local authorities about keeping chickens and building a coop, chosen a site, and checked with your neighbors . . . it's all downhill from there, right?

Not exactly. You haven't picked out a coop style yet. (Then, of course, there's the whole matter of actually constructing it, but let's not get too far ahead of ourselves.) Based on zoning and neighborhood restrictions, location, utilities, drainage, and desired coop size, you may find that only one style of housing will work for your unique situation, or you may have your pick of the entire roster. While your site and your flock size certainly help drive the decision on which coop is best for you, you still have more to consider.

One consideration is pure aesthetics. How do you want the coop to look? Do you want it to match the architecture of your house? Are you hoping it blends seamlessly into the landscape? Would you like it to *look* like a chicken coop, to proudly proclaim your new hobby to the world at large, or would you rather have a subtle structure that could be mistaken for a children's playhouse? It'll be sitting in your backyard, right there in plain view every time you glance out your windows or step outside. You probably don't want to select something that doesn't look good to you.

There's also the DIY factor to take into account. Theoretically, you picked up this book and are reading it because you are at least contemplating getting out there with some lumber and tools and building this thing yourself. Maybe you're a professional handyman who does this sort of thing all day long and can construct any one of the coops we present in his sleep. Great; read on.

Maybe, you're a novice when it comes to carpentry, tools, and such. Perhaps this coop will be the first time you've ever attempted to build anything. If that's the case, you'll want to steer toward the smaller, simpler coops.

The chapters that follow go into detail about everything from tools to materials to basic building techniques. But make no mistake, your self-confidence and overall skill level can (and must) be a consideration when you put your finger on a coop and say, "That's the one I'm going to build."

REMEMBER

Building your own chicken coop can (and should) be a fun, enjoyable, and rewarding way for you to get even more out of your total chicken-keeping experience. But if you don't know a Skilsaw from a socket set (and don't really want to), a number of reputable manufacturers of prefabricated coops are happy to ship you everything you need, with premeasured and precut parts, little bags of fasteners, and a set of directions that make putting a coop together almost as easy as assembling a store-bought bookshelf. Some will even build the coop for you and deliver it in one piece, ready to plop on the ground and move chickens into. Surely your area has plenty of qualified professional contractors and builders who would be more than happy to erect a chicken coop for a new client. Maybe you have a handy neighbor or cousin you could cut a deal with. We hope you decide to build your own coop, and we think this book will help make that practical. But in any case, you shouldn't let the hammer-and-nails element burst your I-want-to-raise-chickens-in-the-backyard bubble.

Ready to check out the coops? The following sections offer a quick thumbnail sketch of a few popular types. These aren't the only chicken coop styles out there, but they represent the most commonly seen coops and are all DIY-friendly. In Part 3, we present five coops, with detailed plans, complete materials lists, and step-by-step instructions.

A-frames and hoops

Taking the same concept in two geometrically-different directions, A-frame coops and hoop coops are among the simplest coops you'll typically find. They're great for small yards where space is at a premium and large flocks aren't practical. They're lightweight and easy to move from place to place in the yard. (Just don't move the coop with the chickens hanging out in the run!)

Because of their small size and basic design, these coop styles require a minimum of materials and know-how to construct. They are, therefore, probably the least expensive coop types to build and great examples of what a low-cost hobby raising chickens can be. Here are a few details on both types:

>> **A-frames:** An A-frame (see Figure 2-4) is long and triangular in shape. Two sides lean against one another and come to a point, serving as both the walls and the roof of the structure. Most often, a solid, covered shelter portion is connected to a fenced run portion in one long unit. While the shelter generally has a floor, the run doesn't — it simply rests on the ground. Usually, both the shelter and the run have a door or hatch that's accessible from outside the coop.

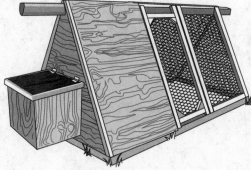

FIGURE 2-4:
An A-frame coop gets an "A" for small-scale simplicity and efficient design.

>> **Hoops:** A hoop coop (see Figure 2-5) is typically rounded and tunnel-shaped. Flexible piping (like PVC) often makes up the frame, with some kind of bendable fencing material covering it. A simple tarp or solid fabric can be draped over a portion of the hoop to make a decent shelter. Many chicken owners with larger coops keep a hoop handy to use as temporary housing during the main coop's clean-out day. A small hoop coop can be very lightweight and may need to be anchored to the ground to keep it from tipping over in high winds.

FIGURE 2-5:
A hoop coop's well-rounded design is a cinch to assemble.

Chicken tractors

No, John Deere has not come out with a miniature model that allows you to stay inside and watch the big game while your feathered friends saddle up and till the back 40. A chicken *tractor* (refer to Figure 2-6) is a type of housing that's meant to be moved from place to place. (A-frames and hoops, described in the preceding section, can be considered a subset of the tractor family.)

A tractor coop typically features a roof and four walls, but no bottom (except in the shelter). This allows the chickens to work the soil and fertilize different patches of ground, depending on where the chicken-keeper places the tractor.

To facilitate moving the coop around the property frequently, most tractors are built either with wheels (refer to Figure 2-6a) or on heavy lumber skids (refer to Figure 2-6b). Smaller tractors can be dragged by the chicken-owner (or the chicken-owner's oh-so-helpful-and-energetic kids), while larger, heavier tractors may require a real tractor to do the pulling.

FIGURE 2-6:
A chicken tractor is meant to be moved around on wheels or skids.

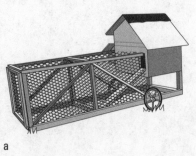

a

b

A tractor-style coop is still fairly economical, thanks to its relatively small size and materials list. Some tractor designs, however, include a real shingled roof which, as discussed in Chapters 7 and 8, can add a level of carpentry that makes some DIYers nervous.

WARNING

Tractor coops built without a side door or top hatch must be lifted to catch chickens. And the chances of your entire flock sitting patient and still while you lift one end of the tractor are smaller than Snoop Dogg and Paris Hilton's chances of winning Oscars for a *West Side Story* remake.

All-in-one coops

Many owners like how A-frames, hoops, and tractors incorporate the shelter and run into one unit, but don't need the coop to be portable and want it to be big enough for them to enter themselves. Welcome to the all-in-one coop.

As shown in Figure 2-7, an *all-in-one* coop usually features a shelter and a run under one tall roof that covers everything and provides human access to the run but not the shelter. This access — for egg-collecting, feeding, cleaning, and chicken-gathering — makes this an attractive coop style where carefree maintenance is a goal.

WARNING

All-in-one coops start to get larger than the styles previously outlined, with real stud walls, tall support posts, and an actual roof that includes rafters and such. The all-in-one coop is a big structure, tall enough for a human caretaker to enter, so the materials list can be lengthy. Building this coop style may require a good bit of skill and probably isn't a one-weekend kind of project.

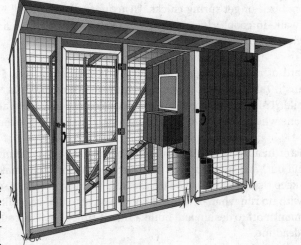

FIGURE 2-7:
An all-in-one coop puts the shelter and run under one roof.

Walk-in coops

If you have the property to pull it off and the skills to slap one together, a walk-in coop (as seen in Figure 2-8) is generally considered the primo option for backyard chicken-keepers. Material cost, space, and build time notwithstanding, a walk-in coop is unquestionably the most comfortable housing choice for chickens as well as their keepers.

FIGURE 2-8:
A walk-in coop
is generally
considered the
"coop de ville"
of chicken
housing.

As the name says, a walk-in coop is large enough for you to stand upright in. (Think of it as a full-size toolshed with chickens living in it.) This obviously makes maintenance, cleaning, and all other chicken chores much easier on you, and it gives your flock a lot of room to move about inside their shelter.

REMEMBER The primary distinguishing feature of a walk-in coop is a heavy floor that allows you, the caretaker, to enter the shelter. For the floor to support you adequately, you must use the right materials and build a proper support system underneath. We deal with these issues in Chapters 4, 6, and 7.

The larger size also means you can increase the size of your flock more easily (or just start out with more chickens) than if you had a smaller coop. That starter flock may seem like a big responsibility at first, but most new caretakers find that they love raising backyard chickens and can't wait to buy more birds or get spring chicks. "The more, the merrier" is an easy philosophy to adopt with a walk-in coop; it's generally not even an option with a tiny A-frame (unless you build another one).

But even the gold standard of chicken coops comes with a drawback or two. For starters, a walk-in is big. That means it takes up a lot of space, requires a lot of material, and demands a good deal of skill to build. This kind of coop will almost certainly take several weekends to construct, even for someone who knows what he's doing and invites friends over to help.

WARNING Walk-in coops seldom factor in a run. The outdoor run is almost always a completely separate entity, to be designed, laid out, and incorporated however you see fit. That makes it like two fairly sizable projects, so keep that in mind as you decide. And because most city-dwellers can't realistically have a coop with no run whatsoever, it's not practical to build a walk-in, move the birds in, and then take a month off to design and build a run. If time is in short supply, this one may be tough to do on a deadline.

TIP If you live out in the country, where you have a lot of land and where neighbors (and their dogs) and predators aren't really an issue, you don't technically *need* an enclosed chicken run. In these situations, there's no rule that says you can't erect a nice walk-in coop and then just allow your flock free-range access, so they can come and go anywhere their little chicken hearts desire. If they know where "home" is, they'll come back at dusk and put themselves to bed. (Just remember to go out and lock the coop at night to keep them safe from four-legged midnight snackers.) Free-range chickens come with both pros and cons, but if you have room for them to roam and want to give it a shot without a dedicated run enclosure, I have great news: We just saved you a bunch of money on your coop construction costs.

TIP

Most backyard chicken-keepers eventually end up with a walk-in coop, even if they start out with a smaller coop style. But it can be a sizable investment, especially if you're just getting started and aren't 100 percent sure that raising chickens is truly in your blood. Don't be afraid to begin your hen hobby with a compact, low-cost A-frame, hoop, or tractor to test the waters. You can always graduate to a walk-in when your time and budget allow, and you'll have the smaller coop to use as a backup, for temporary housing during cleaning or to quarantine sick birds from the others — or even just to give your problem chicks a timeout when they've been bad little birdies.

Chapter **3**

Gathering Your Gear

W hether you're building the simplest A-frame hut or constructing a scale replica of the Taj Mahal, turning a pile of raw materials into the coop of your chickens' dreams requires the right tools as well as the skills to use them.

But don't worry; you don't have to be a master craftsman to build a house that your flock will be proud to call home. As long as you're comfortable working with just a few basic tools and willing to put in a little time and effort, building your own chicken coop is a great DIY project that, at the end of the day (or weekend . . . or several weekends . . .), gives you a fantastic sense of accomplishment and your chickens a fantastic roof over their heads.

This chapter runs down the tools you need to have on hand (Chapter 5 delves into how to use basic tools effectively and safely). All the tools mentioned can be found at your local hardware store or home improvement center. Many can also be rented for short-term use.

REMEMBER

If you're a serious woodworker or just an avid collector of cool tools, you'll no doubt find an excuse to drag every single piece of equipment you own out of your workshop while building your coop. But coop construction doesn't necessarily require a huge arsenal of horsepower-heavy machinery. True, a well-stocked workshop can certainly make the building process more fun, and a few certain specialty tools make specific steps of the build easier. But they're not a must. You can construct a perfectly workable coop with nothing more than the following everyday tools, most of which are probably rattling around in a toolbox in your garage or basement right now:

>> A tape measure

>> A circular saw

>> A hammer

>> A drill (used as a screw gun)

Putting Safety First with Essential Equipment

WARNING

Before you cut your first 2x4 or hammer home a single nail, spending a moment on safety is essential. Any tool can do bodily harm if used improperly, and many tools are downright dangerous just by their very nature, even in the skilled hands of a pro. No chicken coop is worth a trip to the emergency room. Using a little common sense and taking a few simple precautions should be the first step in any DIY project.

REMEMBER

Here are a few basic pieces of safety gear that no toolbox should be without:

>> **Earplugs:** Power tools are loud, and you're likely to be in very close proximity to them while you build your coop. Using even an inexpensive pair of throwaway foam earplugs protects your eardrums from potential permanent damage. (If only you'd had them that time you were in the front row at the AC/DC concert. . . .)

>> **Gloves:** Rawhide leather, form-fitting nylon and neoprene, even pretty pink polyester — it doesn't really matter. A good pair of work gloves gives you a better grip on your tools; protects against scrapes, nicks, and cuts; and helps prevent painful blisters.

WARNING

Make sure the gloves you wear fit properly! Gloves that are too big or too loose can easily get caught in a power saw's spinning blade and prove to be more dangerous than not wearing gloves at all.

>> **Goggles:** Forget those clunky, elastic-banded geek glasses from high school chemistry class. Safety goggles can be as comfortable as your designer sunglasses, and just as stylish. They also keep stray bits and pieces of dust and flying material from putting your eye out, which would be a bad way to end a good build day.

>> **A tool belt:** No one ever mistook a tool belt for a chic fashion accessory. But what the tool belt lacks in style, it makes up for in convenience, practicality, and safety. A clean work-site is a safe one, and all of those pouches, pockets, and loops keep your hand tools and fasteners within safe reach. You can't accidentally trip over your hammer if it's hanging from your waist.

Digging Up Dirt on Garden Tools

Depending on the desired location of your coop, using garden tools may not be necessary. Many chicken-owners place their coop on a plot of flat, grassy lawn that seems custom-made for chickens. But in some yards, you have to clear out a spot. This may be as simple as digging up a few shrubs, or as backbreaking as removing old tree roots and busting up some rocky ground.

Some tools you may find useful for prepping your site are

» **A rake:** For clearing away wet and packed leaf litter or mulch, a rake is as easy as it gets. A bow rake with metal tines has more pulling power than a plastic leaf rake.

» **A shovel:** A flat-bladed shovel or garden spade is an effective tool for slicing underneath sod, should you need to do that. A shovel with a pointed blade is the way to go for digging up shrubs or plants to make room for your coop.

» **A mattock:** If your coop location is inundated with close-to-the-surface tree roots, a *mattock* (see Figure 3-1) is a good weapon of choice. On one end is a small blade similar to that of an axe, perfect for chopping through roots. The other end features a broad, chisel-like blade that can be used for prying or trenching.

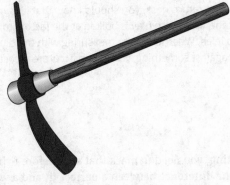

FIGURE 3-1: A mattock makes short work of stubborn tree roots and shallow rock.

TIP

If clearing the site for your particular coop requires anything more than these rudimentary landscaping tools (like a backhoe, for example), you may want to rethink the location of your coop. Finding a new spot is probably easier than taking on full-scale excavation as your very first step. Check out Chapter 2 for guidelines on selecting a spot for your coop and Chapter 6 for the ins and outs of preparing your site.

Measuring and Marking Lumber for Your Coop

If you're building a coop from a prefab kit, most (if not all) of the lumber will come in handy, precut, ready-to-assemble lengths. If you're constructing your own coop, it's up to you to turn those long pieces of lumber into exactly what you need. You'll be spending a lot of time with the measuring and marking tools described in the following sections. (Flip to Chapter 4 for the basics of shopping for lumber.)

Measuring up tape measures

When Noah built the ark, he measured out his lumber using not feet and inches, but an ancient unit of measurement called the *cubit.* That's the distance from the elbow to the tip of the middle finger. But unless you're kicking it *really* old school, using cubits probably isn't how you want to measure out your lumber when building a coop.

As a matter of fact, the modern tape measure is arguably the single most important tool you'll use during the build. And while it may seem like the most straightforward gizmo of them all, some features warrant a closer look:

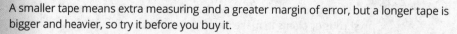

TIP

>> **Length:** Tape measures come in different sizes, but 10-, 12-, and 25-foot-long tapes are the most common. You don't need anything longer than that for typical coop construction.

>> A smaller tape means extra measuring and a greater margin of error, but a longer tape is bigger and heavier, so try it before you buy it.

>> **Lock:** Most tape measures have some sort of locking mechanism that holds the tape at any desired length. This is extremely handy for transferring the same measurement over and over to several pieces of lumber. Some models employ a thumb-controlled lever, while other tapes automatically lock in place and require a manual release. Which one you use is a matter of personal preference; they work the same.

>> **Edge clip:** The metal hook on the end of a tape measure is called the *edge clip.* And before you march back to the hardware store's return desk, you should know that it's *supposed to* wiggle a little bit on the end of the tape. This loose play is built in at the factory to compensate for the thickness of the clip itself. Whether you're measuring with the clip hooked over something or pushed tight up against something, you get an accurate measurement either way.

Making your mark

Once you've measured a board for cutting, you need to mark that spot before firing up a power saw. But *how* you mark it may mean the difference between a perfect fit and a wasted piece of wood. Consider these marking tools:

>> **A carpenter's pencil:** A pencil is a pencil, right? Well, kind of. A carpenter's pencil (see Figure 3-2) is oversized, making it easier to grab and manipulate while wearing work gloves. Even better, it's flat, so it doesn't roll away when you set it down.

FIGURE 3-2:
A carpenter's pencil stays right where you put it.

>> **A chalk line:** When you're making a long cut, like across a full sheet of plywood, a chalk line (see Figure 3-3) is a quick way to mark the entire length in a snap — literally. String is coiled up inside a container filled with chalk dust. When the string is pulled tight and snapped against a surface, it leaves a brightly colored line to refer to as needed.

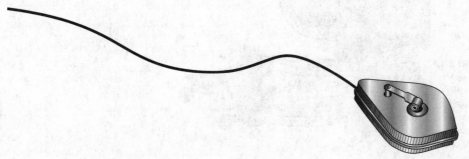

FIGURE 3-3:
A chalk line is a bright idea for marking a long cut-line.

>> **A self-marking tape measure:** File this away as one of those "I wish I had thought of that" ideas. Tape measures are now on the market that have a marking device built into the bottom of the housing. Once you find your exact measurement, you simply touch the tape measure to the workpiece. Is this advancement in tape measure technology worth a special trip to the home improvement store? Probably not for most folks, but if you have to buy a new tape measure and are habitually losing your pencil, it may be worth considering.

Sizing Up Saws and Supports

Walk down the saw aisle of the hardware store, and you may be overwhelmed by the obscene assortment of sawing implements available to anyone with a credit card. It's like a slasher flick's prop room: chainsaws, jigsaws, reciprocating saws, saws with blades that stick up menacingly out of shiny tabletops, saws with intricately-carved handles and grips, saws for wood, saws for metal, saws for plastic, even saws with laser guidance systems.

For most 'round-the-house projects, though, just a few different saws can almost certainly get the job done. For a chicken coop build, you can probably get away with just one or two of these basic saw types:

>> **Circular saw:** Commonly called a *Skilsaw* (after the company that invented it), a circular saw is by far the one used most frequently by do-it-yourselfers. It's lightweight, versatile, and easy to control. Cutting a truly straight line, though, requires either a very steady hand or the use of some kind of straightedge. Blades can be swapped out depending on the material you're cutting through.

>> **Miter saw:** A *miter saw* (see Figure 3-4), or "chop saw," generally features a larger blade than a handheld circular saw, so it can cut through bulkier pieces of lumber. It's a stationary tool, meaning the piece to be cut has to come to the saw. A powerful motor, a metal fence and bed to place the lumber against, plus factory-set stops at 90 degrees, 45 degrees, and common angles in between allow you to cut through lumber as easily as slicing through a stick of butter.

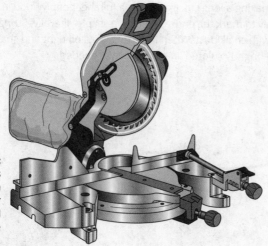

FIGURE 3-4:
A miter saw is practically a must-have for fast and accurate straight cuts.

TIP

Miter saws are not inexpensive, so you may not necessarily want to invest in one of your own if you're not a big DIYer. Consider marking all your lumber first and then renting one for a day to tackle all the cuts in a single afternoon. They're so fast and precise, it's money well spent.

» **Handsaw:** Yes, Virginia, there are still handsaws out there. But unless manually pushing and pulling a blade through countless 2x4s — or worse yet, 4x4s — is part of some hard-core lumberjack workout regimen you're trying, a handsaw shouldn't be your tool of choice for the bulk of this project. Having one handy is recommended, though, because it does a great job of finishing off cuts that may be too deep for your circular saw or miter saw to complete.

» **Table saw:** A table saw (see Figure 3-5) can be a true luxury when it comes to cutting or ripping sheet goods. (*Ripping* is simply fancy woodworkers' slang for sawing a long piece of wood lengthwise, or with the grain of the wood.) Maneuvering a handheld circular saw down the length of an 8-foot-long piece of plywood and keeping anything resembling a straight line is a real challenge for any craftsman. It's far easier to feed that sheet of plywood into a table saw, gently sliding the wood across the tabletop surface and letting the saw's stationary fence worry about keeping the edge straight. (If you're lucky enough to have a table saw or you're considering buying one, check out the nearby sidebar "Accessorizing your table saw" for accompanying gear you need.)

» **Jigsaw:** A handheld jigsaw is typically reserved for making curved cuts in a variety of materials, but this tool can be quite adept at ripping sheet goods as well. With a decent straightedge as a guide, cutting a full sheet of plywood with a jigsaw is fairly easy. It takes longer than using a table saw or circular saw, but many people find that controlling and maneuvering the lightweight jigsaw over a long cut is far easier than manhandling the larger saw types.

ACCESSORIZING YOUR TABLE SAW

Fashion models know that the accessories make the outfit. A little black dress is nice, but the right jewelry, purse, and shoes take it from *bor-ing* to *ba-boom!* Here's a look at some accessories that all the safest table saws are wearing:

- **Roller stand:** A full sheet of plywood is heavy and unwieldy all by itself. Maintaining control of it as you feed it through a table saw is so difficult that it could be an Olympic sport. A roller stand set to the same height as your table saw and a few feet away gives you an extra set of hands and helps support the piece as it comes off the tabletop.

- **Extensions:** Most table saws can be outfitted with extensions that expand the surface of the tabletop. A combination of extensions and a strategically-placed roller stand or two can make cutting a full sheet of plywood a one-person job.

- **Push sticks/shoes:** Maintaining good leverage on the workpiece can be tricky, especially as you near the end of the sheet. Your hands have nowhere to go but closer and closer, inch by terrifying inch, to that spinning saw blade. A push stick (or the larger push shoe) is used to guide that workpiece through the saw while keeping your fingers at a safe distance. These can be purchased or made from scrap wood. If you use a table saw, this safety accessory should be a nonnegotiable add-on.

FIGURE 3-5: A table saw is tops for cutting sheet goods like plywood.

TIP

In addition to a saw of some kind, enlist a few handy helpers to keep things stable and steady while working with power saws:

- **Sawhorses:** A good set of sawhorses is crucial. They can be set up as a temporary workbench, act as a platform for your miter saw, and keep you working at a comfortable waist height instead of down on the ground.

- **Clamps:** Clamps are the DIYer's best friend. The degree of difficulty in making a clean, straight cut through a piece of wood skyrockets if you don't have the piece secured. Tightly clamping a board to a solid work surface frees up one hand to better guide your saw through the cut.

Putting In Posts

Many coops are built to simply rest on solid ground. This kind of skid-frame construction allows great flexibility for future use. Want to move your small coop to another location? Just grab some strong-backed buddies and heave-ho (after temporarily evicting the chickens to a safe place, of course). Some coops are built on wheels, specifically designed to be towed to different areas in the yard to allow the chickens to turn and fertilize the soil. (See Chapter 2 for more about these types of coops.)

But for many coop-owners, it makes sense to build a coop as a permanent fixture in the landscape. If your ideal coop location is on a slope, for example, constructing it on stilts that are anchored in the earth may be your only option. And if you want to use permanent support posts, you need the tools in the following sections. (See Chapter 6 for details on putting in posts.)

Digging postholes

To get your coop up off the ground, you may need to install *footings:* weight-bearing supports that are buried in the ground. For a simple, modest-sized coop like the ones we describe in this book, you can get away with four footings — one on each corner.

Typically, footings are cylindrical in shape and made of poured concrete, with a timber post encased inside and sticking up out of the ground to build off of. But it all starts with digging a big hole. And that means . . . more tools:

>> **Posthole digger:** The lowest-tech way to dig postholes is with the unimaginatively-named *posthole digger* (see Figure 3-6). Also nicknamed a "clamshell digger," it's a bit like two shovels facing each other and fastened together, like a set of giant landscape tongs.

FIGURE 3-6:
A posthole digger won't win a race, but it's effective for digging footings.

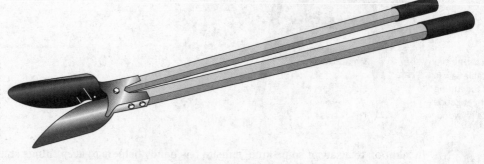

>> **Power auger:** Not digging the idea of excavating those footing holes by hand? There's a revved-up option that may cut the time significantly. A *power auger* (see Figure 3-7) is a massive, gasoline-powered drill made for digging deep, round holes. You can find one at most tool rental shops, along with spiral-shaped drills in a variety of widths and lengths.

FIGURE 3-7:
A power auger brings horsepower to a hole-digging party.

WARNING

The digging itself goes quickly with an auger, but time is obviously involved in going to the rental shop, getting the tool home, cleaning it when you're finished, and returning it to the shop. For a couple of holes, it may not be worth it. And for many augers, you need a willing partner to help you maneuver the auger's double handlebars and handle the sheer weight of the machine. Chapter 6 has the ins and outs of using an auger.

Setting the posts

When you're ready to physically set a post in one of your spiffy new postholes, you need to get it standing perfectly straight — *plumb* — and secure it in that position before adding concrete. These tools can help:

>> **Using a post level:** A *post level* (see Figure 3-8) is a handy, yet hands-free, device for making sure that your post doesn't resemble the Leaning Tower of Pisa. It works like a carpenter's level (described later in the section "Just level with me"), with a few key advantages:

- The post level's "L" shape allows it to sit snugly against the corner of a post, with strategically placed bubble vials within easy view.

- A rubber band wraps around the post and holds it fast, freeing your hands to adjust the post until the bubbles read level and plumb.

>> **Using braces:** Assuming you don't want to hold that post perfectly plumb until the concrete is mixed, poured in, and completely hardened, you need to find a way to secure the post in the plumb position while you continue working. Braces are your answer. Because they're just temporary tools, braces are usually made on-site using scrap wood or lengths of inexpensive lumber. Check out Chapter 6 for a quick step-by-step, how-to guide to making your own braces. For now, plan on buying some extra 1x2s, about two 8-foot lengths for each post you have.

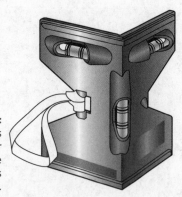

FIGURE 3-8:
A post level is essential for making sure your posts are plumb.

Mixing and pouring concrete

Concrete is hard. Mixing up and pouring a batch of it doesn't have to be. There are several methods you can use, each with its own list of stuff you need to do the job. But as long as you have these basics, you're good to go:

» **A wheelbarrow:** First things first: It's called a *wheelbarrow,* not a *wheelbarrel.* And it comes in extremely handy for many phases of the coop build. You can use it to shuttle tools and materials back and forth to the build site. Cleanup is a breeze when you use it as a portable trash receptacle for those bent nails and scrap corners of lumber. And, it's the ideal vessel for mixing bags of concrete and pouring it into postholes.

TIP

If you're shopping for a new wheelbarrow, consider one with a pair of front wheels instead of a single tire. Every landscaper on the planet has accidentally dumped a load of something in the wrong spot trying to navigate a heavy, single-wheeled wheelbarrow through a tight turn or over uneven ground. The two-wheeled models are much easier to balance and handle, even when they're fully loaded.

» **A garden hoe:** It's the perfect tool for mixing concrete, although you can get away with just about any other long-handled implement: a shovel, a rake, even a piece of scrap lumber, if you're in a bind. A mortar hoe (see Figure 3-9) is perhaps easiest of all, thanks to holes in the blade that help the ingredients mix faster.

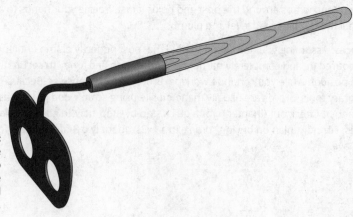

FIGURE 3-9:
A mortar hoe is designed to mix concrete quickly and thoroughly.

There are several other high-tech ways to mix concrete, including some plug-in power options; clever, specialized gadgets; and even alternatives that require no mixing whatsoever! We deal with these methods in Chapter 6.

Framing Your Coop

When you get right down to it, building a chicken coop (or just about anything else) is, at its core, just putting a bunch of pieces together. Taking those 2x4s and fastening them to one another to create the outline — the frame — of your chicken coop is called *framing*. Ah, but *what you use* to put those pieces together — that makes all the difference.

As with just about every other tool choice, you have several options here. Each has its pros and cons: Each has some things it's good at, and each has some things it doesn't do quite so well. In all likelihood, you'll use one of the following tools (or two, or maybe even all three, at different phases of the build) to do your framing. (Flip to Chapter 4 for information on the nails and screws you need for your coop build.)

Honing in on hammers

Do hammers require an explanation? Well, yes. All hammers may serve the same general function, but the one that's buried in your junk drawer right now may not be what you should use to build a chicken coop. There's a specialty hammer for practically every purpose under the sun (sledgehammer, ball peen hammer, brick hammer, rock hammer, drywall hammer, MC Hammer . . . oh, wait), but for assembling your coop, you need a basic carpenter's hammer, also called a claw hammer. Here's what you need to consider:

TIP

>> **Weight:** The heavier the hammer, the more force you bring with each swing. Fewer swings equal faster nailing. But a heavier hammer can be harder to control and can tire you out more quickly.

It's a good idea to pick up a hammer and give it a few test swings in the store (watch out behind you!) to see whether it feels good in your hand. Now imagine swinging it with force, repeatedly. Over and over. All day long. But beware: A hammer that's too light may not have enough oomph to fully sink your nails. For this kind of project, don't go with any hammer that weighs less than 16 ounces.

>> **Claw:** The two-pronged claw at one end of the hammer is most often used for pulling out bent nails. On what's called a *curved claw hammer,* the claw, as you may have guessed, curves downward dramatically. If the claw is only slightly angled, what you have is a *straight claw hammer* (even though it's not truly straight). Curved claws provide more leverage for pulling out nails, while straight claws do a better job of prying nailed boards apart. Pros like to have one of each at the ready, but if you're picking out just one hammer for your build, you'll probably get more use out of a curved claw model.

>> **Face:** The *face* is the surface opposite the claw that actually strikes the nail to drive it. This is the business end of the hammer. Most general-use claw hammers have a smooth face. A true *framing hammer* (see Figure 3-10), however, has a rough, waffle-like texture on its face. This *milled* face helps maintain solid contact with a typical framing nail, preventing

potentially-dangerous glancing blows. Framing hammers usually feature heavy heads, straight claws, and extra-long handles.

While the framing hammer's milled face can help you avoid off-the-mark swings, it's not foolproof. There's no guarantee that you won't bend a few nails or smash your thumb if your technique is poor. Also, be aware that the milled face leaves that same dimpled pattern on the lumber as you drive the nail home. Not a big deal if it'll be covered by siding or plywood sheathing, but on a surface where appearances matter, a framing hammer may do more harm than good.

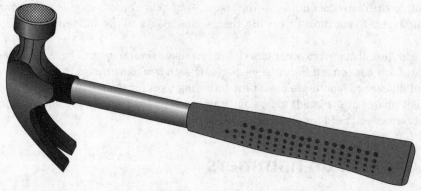

FIGURE 3-10: The face of a framing hammer helps drive nails solidly but can mar your lumber.

Thinking about using a nail gun instead? (See more on nail guns in the section that follows.) Sure, a nail gun is faster, but remember that using one also requires an air compressor and hose or expensive fuel cells, plus special nails that cost a lot more than standard bulk nails. A hammer is a low-cost tool that needs no setup time or other accessories, and even provides a bit of an arm workout during a long build day!

Nailing it with a nail gun

Any visit to an active construction site is usually accompanied by a soundtrack that could have been lifted straight from a World War II movie. That rapid-fire *pop-pop-pop-pop* sound you hear is the professional builder's favorite tool, the nail gun. This bad boy can be powered by an air compressor (in which case it's referred to as a *pneumatic nailer*) or by self-contained fuel cells (in which case it's called a *power nailer*). Nail guns sink nails of any length into almost any material with just the gentle squeeze of a trigger.

A pneumatic nailer is perhaps the single biggest timesaving tool you can employ in your entire chicken coop project. If you don't own one, don't know a neighbor who will loan you one, and don't believe that Santa will bring you one this year, renting one is a low-cost way to reap all of this tool's benefits. If you're building a large coop, you may find that renting a nail gun and the needed accessories pays for itself within minutes of firing it up on the jobsite. And, pneumatic nailers are totally fun to use.

Of course, there are many different types of pneumatic nailers, and each one is specifically suited for a different application. You may decide to use different nailers at various phases of the build, as follows:

- » **Framing nailer:** Probably the closest thing there is to an all-purpose nail gun, a framing nailer typically uses nails that are between 2 and 3½ inches long. It's designed for assembling stud walls, subfloors, rafters and joists, and so on.

- » **Roofing nailer:** Engineered for fastening shingles, a roofing nailer dispenses specially-made roofing nails from a long coil. The roll of nails is loaded into an on-board canister.

- » **Finish nailer/brad nailer:** Although it looks like just a small-scale version of a framing nailer, a finish nailer drives smaller, thinner nails than its larger cousin. While you may be tempted by its convenient size and smaller price tag, a finish nailer's headless nails are used for things like molding and trim-work, and therefore are probably too light-duty to do much good on a coop build. A brad nailer is even smaller and really useful only for fine woodworking or craft projects.

- » **Palm nailer:** About the size of a baseball and usually weighing just a pound or two, a palm nailer is like a nail gun without the gun. A magnetic recess holds a nail head and when pushed, the tool quickly jackhammers the nail into place. It uses whatever regular bulk nails you choose and is especially adept at driving nails into tight spaces.

For most DIYers, a nailer really makes sense only on a large coop build where hundreds and hundreds of nails need to be driven. The actual nailing may be faster with a gun, but the setup of the compressor, expense of the accessories, and added cost of the specialized nails often outweigh the tool's benefits if all you're building is a small-scale coop.

REMEMBER

If you decide to go the nail gun route, be sure to buy the nails that fit your particular tool and usage. If you're using a power nailer, buy the appropriate fuel cells. For pneumatic models, one air compressor can run all nailers.

Pressing a screw gun into service

Okay, a screw gun is really a drill. But the word "drill" implies that you're boring small holes in your work surface. Outfit that same tool with a screwdriver bit, and you have a drill/driver — a screw gun — an effective and easy-to-use alternative to swinging a hammer all day. Using screws versus nails for a project like this is its own separate debate, one we tackle in more detail in Chapter 4. For now, keep these things in mind about screw guns:

TIP

- » **Cut the cord:** Yes, there are still drills out there that you have to plug in to an electrical outlet. And sure, they work just fine. But for a job like this, you'll appreciate the ability to be cord-free. You'll be up on a ladder, down on the ground, back and forth, inside and outside, maybe even under the coop as it comes together. Dragging an extension cord everywhere you go is, well, a drag. And a safety hazard on the jobsite.

- » **Talk about torque:** Most screw guns have a collar around the barrel with numbers and a selector arrow. Those numbers show the setting for the drill's *torque,* or twisting power. The higher the number, the more torque used to drive the screw. Higher torque settings may come in handy when driving screws into especially hard lumber.

TIP

- » **Keep it charged:** If your cordless screw gun doesn't come with an extra battery, buy one. Nothing is more frustrating than having an entire build day grind to a sudden halt because your lone battery ran out of juice. Bring the battery charger to the worksite, keep it plugged in, and always have a battery charging. That way, you're always working.

Leveling and Squaring As You Build

As you build your coop, you want to check often to make sure that things are *level* (perfectly horizontal), *plumb* (perfectly straight up and down), and *square* (all right angles are a perfect 90 degrees). Granted, these are chickens, and they don't require a home that's built to precise NASA specifications, but keeping things level, plumb, and square as you build, with the help of the tools in the following sections, makes for a much easier project and, in the long run, a much sturdier coop.

Just level with me

To make sure what you're building is level, these two useful tools should be in every toolbox and used frequently during the build:

TIP

>> **Carpenter's level:** Also called a "spirit level" by some seasoned pros, this long, beam-shaped tool has clear vials encased in it. Inside each vial is a colored liquid — usually ethanol or some other "spirit" — and an all-important bubble that makes the level such an indispensable tool. Most levels hold multiple vials, one to be used vertically to determine whether up-and-down structures are plumb as well as one that's used horizontally to ensure that all things running crosswise are level.

> You can find levels in many different lengths, from an 8-inch *torpedo level* (which fits nicely in a tool belt or your back pocket) to huge 8-foot models. On a coop build, anything longer than a 4-foot level is probably overdoing it; leave those for professional contractors.

>> **Line level:** To check a span that's longer than your carpenter's level, a line level or string level (see Figure 3-11) is astonishingly simple. It's essentially a single-bubble vial, similar to what you'd find in a carpenter's level, made to hang by a string. Now you can check for level across any length, limited only by how much string line you have.

FIGURE 3-11:
A line level is ideal for making sure post tops are level with one another.

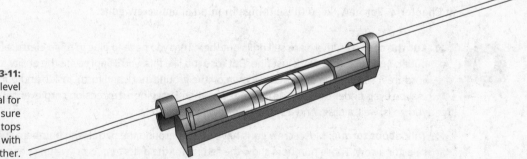

It's hip to be square

When two pieces of lumber are fastened together at a right angle, you might assume that they meet at a perfect 90-degree angle. But remember what happens when you assume? (See, Mom? I *was* listening.) If that angle is just a few degrees out of "square," and you just keep on building, things can really get out of whack the further you go.

Luckily, you don't need to drag any long-forgotten geometry formulas out of your mental way-back machine as you build. To check for squareness, just use a square — a simple tool that establishes a true right angle at a glance. In addition to helping you make sure that two surfaces meet at precisely 90 degrees, a square can help you scribe a perfectly straight line before making an important cut. (And really, aren't they *all* important?)

There are several types of squares, but these are common and user-friendly:

» **Speed square:** Despite its seemingly-obvious name, a *speed square* (see Figure 3-12) is actually a *triangular* piece of metal or high-impact plastic. This deceptively simple-looking tool performs dozens of complicated carpentry functions, but it's most often used to lay out either a 90- or 45-degree angle. Its raised lip fits snugly against a board for easy scribing or even as a sturdy cutting guide for your circular saw. (We discuss circular saws earlier in this chapter.)

» **Steel square:** This long piece of L-shaped steel is an excellent tool for marking longer straight lines. You can also use its perfect 90-degree angle to double-check your work once you start building.

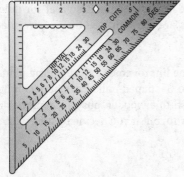

FIGURE 3-12:
Need a straightedge fast? Grab your speed square.

Working with Wire

Almost every coop features some sort of wire mesh, used to protect precious chicks from hungry, clever, and persistent predators. While we take a closer look in Chapter 4 at the various types of wire mesh products you can use on a coop or run, suffice it to say that the giant roll of whatever you buy from the hardware store has to be cut down to size to fit your particular needs and fastened securely. The following tools do these jobs well.

Cutting wire

TIP

As you build your coop and raise your chickens, you'll find a multitude of ways to use sturdy wire mesh, from plugging holes to other quick fixes. Always having a roll on hand is a good idea; so is always having a tool that you know will cut it. While most household toolboxes include some sort of wire-cutting implement, it's usually a light-duty pair of pliers. Even if it manages to cut through your wire, it may not do so easily. Save yourself a headache and a pair of sore hands; get a good pair of tin snips (see Figure 3-13). These monster scissors slice through heavy-gauge wire with ease.

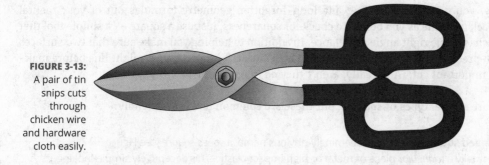

FIGURE 3-13:
A pair of tin snips cuts through chicken wire and hardware cloth easily.

Fastening wire

After your wire is cut, you need to securely fasten it in place. Most household staple guns (the kind you might use for arts and crafts) are insufficient for fastening wire. They usually aren't beefy enough to drive a staple into lumber and can leave you with fencing that's easily pried open by the sharp claws of a chicken thief.

At first glance, a hammer tacker may look like the answer to your wire-fastening needs. If a staple gun and a hammer ran off to Vegas and got married, the result might be this tool: you swing it like a hammer; it drives a staple where it lands. While it's great for quickly securing insulation, roofing felt, or shingles, the staples are too lightweight for wire that's meant to keep strong and nimble predators out of a run.

Buried somewhere in the nail aisle of every home improvement store on the planet is a selection of poultry staples, or fencing staples. These U-shaped nails have barbs on both ends and are designed specifically for fastening wire mesh to a wooden post or framework. We discuss the ins and outs of using these staples in Chapter 10, but if you decide to stick with staples, your hammer may be all you need.

Checking Out Other Miscellaneous Tools

REMEMBER

This chapter is a list of the most commonly used tools for building a coop. As the corporate lawyers like to say, though, your results may vary. If you find a tool that you like better than those we describe here, by all means, use it. You'll no doubt come up with your own tricks and techniques for executing the steps of building your own particular coop, and you'll discover a slew of tools that work well for you in your unique situation. Finding a way to get the job done is what building anything is all about.

As is the case on any job, there are stray bits and pieces of gear that may or may not come in handy. There's no guarantee that each of these tools will be called on, but it's better to have them and not need them than to need them and not have them. Having these extras on-hand may save you some time, some money, or at the very least, your bazillionth trip to the hardware store.

>> **Extension cords and a power strip:** You can't have too many extension cords. You'll be powering up saws, a battery charger, work lights, or even a radio. A heavy-duty extension cord and multi-outlet power strip should be the first things that get put out on the jobsite.

>> **A utility knife:** It's good for cutting open bags of concrete mix, slicing through the nylon straps that may be bundling your lumber, even scratch-marking a board if you managed to leave your pencil *waaaaay* over there on the other side of the yard.

Keep extra knife blades handy and change them often. A dull blade is a dangerous blade.

>> **A screwdriver:** You may need to give that screw one tiny half-twist to make it sit flush after driving it with your screw gun. A good old-fashioned screwdriver helps ensure that you don't strip out the head by using more torque than is necessary.

There are lots of different screw types (common types are discussed in Chapter 4), so have a screwdriver on-hand that fits the specific screws you're using. (This may mean having more than one screwdriver in your tool belt.) Some screwdrivers come with interchangeable bits that fit into the handle, so a single tool can turn a variety of screw types.

>> **Drill bits:** Pilot holes can prevent nails or screws from splitting wood. If you need to drive a nail or screw at an odd angle (see Chapter 5), a pilot hole drilled at that angle can help you get it started.

If you think you'll be drilling a lot of pilot holes before driving screws, consider having two drills at your disposal: one just for drilling and one dedicated to sinking screws. This saves you from having to continually swap bits in and out of your lone drill every few minutes.

>> **A ratchet/socket set or wrenches:** If you'll be using hex-head bolts and nuts or lag screws, one or the other is a must-have. They're pretty much the only tools you can use to really tighten down these fasteners.

>> **A sander (or a file/rasp):** Cutting a piece of lumber often leaves a rough edge with jagged little slivers and shards of wood sticking out, just waiting to snag your clothes and leave you with painful splinters as a souvenir of your build. Keep some sort of sanding or filing tool handy; one quick pass over a fresh cut knocks down any razor-sharp burrs.

>> **A cat's paw:** A cat's paw (see Figure 3-14) is a steel prying tool that works like a hammer's claw. Its exceptionally sharp ends can dig any nail out of a piece of lumber, even if it's fully embedded. It takes a nasty bite out of the wood in the process, but if you absolutely, positively have to get that nail out, sometimes there's no other way.

FIGURE 3-14:
A cat's paw is the purr-fect tool for digging nails out of wood.

Chapter **4**

Deciding on Materials

Every meal, whether it's a simple soup-and-sandwich combo or a gourmet, seven-course dinner, starts out as a bunch of ingredients on a supermarket shelf. Similarly, your brand-new chicken coop is out there right now: full lengths of lumber, unopened boxes of nails, sealed stacks of shingles, uncut sheets of plywood. They're spread out between the hardware store and the lumberyard, hiding in your own basement or toolshed, or maybe even sitting unused at a neighbor's house. These are the raw materials you'll use. Only they don't know it yet. Maybe you don't either.

Whether you're starting from scratch with all-new materials or recycling some items from a previous project, you have plenty of choices to make. Some materials are worth their weight in gold as you build; others cause more problems than they solve and should be avoided like the proverbial plague. New or used, fancy or found, this chapter helps make sense of the wide range of building materials that can be used to construct your coop.

Before You Shop: Considering Recycled Materials

By their very nature, chicken-owners are wonderful improvisers who love being self-sufficient and able to adapt to most any circumstance. As a result, chicken coops are often elaborate amalgamations of whatever materials are handy at the time of the build.

Have a stack of old deck lumber taking up space in your garage? Leftover pieces of plywood from a previous project? An unopened stack of shingles from when you had your roof replaced last year? You kept these things for a reason, right? Thinking that maybe someday, you'd find a use for them? Well, welcome to someday.

Using recycled materials is a fabulous way to cut costs on a coop build. Anything you don't have to buy is money you can spend on other things, chicken-related or not. Your flock won't care if its coop window is the annoying one from your living room that you replaced because it was old and outdated. It won't matter to them if their coop is painted with the gallon that you meant to use in the bathroom until you got over your "orange-is-making-a-comeback" phase. Got corrugated roofing panels in two different colors? It's all the same to the chicks. As long as the material in question is sound, there's no reason not to use it. We've seen particularly enterprising chicken owners use wooden pallets as walls. One even turned a junked 1970 Morris Traveller into a fantastically funky coop!

REMEMBER

Using recycled materials often requires you to deviate from normal building practices and techniques. But if you have a material you want to use on your coop, go for it. There's usually a way to incorporate it safely and securely. It may take some trial-and-error, and you may find yourself asking for ideas and advice from everyone you know: buddies, neighbors, the friendly older guy at the hardware store, even fellow chicken-owners on the other side of the world via online community forums. But that's all part of what makes owning chickens and building your own coop so much fun, isn't it?

Lumbering Through Boards for Your Build

Wood is the most common and most popular building material on the planet. It's versatile, inexpensive, and easy to work with, even for the backyard builder. When wood is cut into standardized boards, it's called lumber. And for the vast majority of coops, lumber is the material used most. In the following sections, we describe lumber sizes and types, and we explain how to examine individual boards to find the cream of the crop for your coop.

Sizing boards

The two-by-four (2x4) is as basic as it gets, with the numbers referring to the board's thickness and width in inches, respectively. Similarly, you'll find 2x3s, 2x6s, 2x8s, 4x4s, 1x2s, 1x3s, and a host of other popular board sizes. But note that "2x4" tells you nothing about how *long* the board is. Each size of lumber can be purchased in a range of standard lengths, from 8 feet to 24 feet. (Turning these boards of lumber into studs, joists, and rafters is tackled in Chapter 7.)

REMEMBER

A piece of lumber is universally referred to by its *nominal dimensions,* not its actual ones. That means that a "2x4" is "two-by-four" in name only. A tape measure reveals that it does not, in fact, measure 2 x 4 inches. The actual measurements of a 2x4 are 1½ x 3½ inches. (A 4x4 is 3½ x 3½ inches. A 1x2 really measures ¾ x 1½ inches.) Remember to take these actual *fractional dimensions* into account when building. And when in doubt about a board's true dimensions, always, always, always double-check them with a tape measure before doing something that can't be undone.

If you're building one of our coops (see Part 3 of this book), refer to that coop's specific materials list for an exact count of the number of boards you'll need to buy and the different sizes required. If you're designing your own coop, stick with these generalities when it comes to what size lumber to use where:

>> **2x4/2x3:** The basic 2x4 is a solid bet for simple stud framing of any structure. We also like the 2x3; although sometimes a bit harder to find, it's plenty sturdy and strong enough for your framing needs. But it's narrower by an inch, making it less expensive and lighter. And it saves you some valuable space inside the coop!

>> **2x6/2x8:** Wider than the 2x4, these beefier lumber types are most often used for framing a subfloor in a large walk-in coop, which must support human weight. They're also commonly used in the creation of roof rafters for large structures. In a small coop only accessible to chickens, though, they may qualify as overkill.

>> **4x4:** Most often used as a vertical support post, a 4x4 provides exceptional strength. As far as chicken coops go, you'll likely need this only if you're building a walk-in coop whose floor is to be elevated off the ground, like a deck. (But our "Minimal" coop in Chapter 12 uses them as framing members that eliminate the need for internal studs.)

>> **1x4/1x3/1x2:** Any piece of lumber that's "one-by-something" is pretty thin — ¾ inch, to be exact. This makes it too flimsy to provide any real structural support, but it comes in handy for all kinds of trim-work. You may also find a need to use thin stock like this as internal blocking — internally-fastened pieces that only help hold other pieces together.

Figuring out what type of board to buy

You'll find many kinds of lumber at your home supply center. Here's a quick rundown to help you choose the best type for your application:

>> **Construction lumber:** Construction lumber is what you'll find the most of at the local lumberyard. It's milled from softwoods (usually meaning pine, fir, or some other evergreen) and has the stereotypical whitish-yellow lumber color. It's meant for use indoors or where the lumber itself won't be exposed to moisture or weather.

TIP

National grading rules have been developed to classify construction lumber, resulting in boards that feature a stamp displaying the piece's grade, as well as species, moisture content, and even the mill where it was produced. It's a confusing array of abbreviations and codes that many carpenters don't even bother to learn. In a nutshell, if it's for sale at your local big-box supply center or lumberyard, it's fine for you to use to build a chicken coop.

>> **Pressure-treated lumber:** While construction lumber is intended for indoor use, green-tinted, pressure-treated lumber is made for outdoor exposure. But contrary to popular belief, the chemicals used to pressure-treat lumber aren't designed to combat moisture. Pressure-treated lumber is all about making wood that will get wet unattractive to pests, fungus, and bacteria that would otherwise converge on it like it was a dessert bar. Nevertheless, pressure-treated wood is the way to go for lumber that will be exposed to the elements, because it results in a longer-lasting structure. If you have safety concerns about this kind of wood, check out the sidebar "The people versus pressure-treated lumber" for more information.

TIP

Different classifications of pressure-treated lumber exist for boards that will be used above-ground, on-the-ground, or underground. Consult with a sales associate at the lumberyard to make sure you select the lumber that suits your application.

Pressure-treated lumber costs more than construction lumber. But many DIYers prefer to have just one big pile of lumber to pick through as they build, as opposed to some boards that can be used only for this function, and other boards just for that function. Because the coop is outdoors and may be exposed to high and prolonged levels of moisture and humidity (and occasionally hosed out for cleaning!), many builders decide to use pressure-treated wood exclusively for their coop. Even though they spend more on lumber, they gain a certain amount of peace of mind knowing that the coop itself will last longer.

» **Cedar and redwood:** Cedar and redwood offer chemical-free resistance to rot and insects, thanks to their naturally-occurring tannins and oils. They're beautiful woods, but they can also be quite expensive. You may not find all of the same board sizes at your local home improvement warehouse. While these woods may be a welcome feature for exposed pieces of the coop, using them for hidden framing members or interior structural supports probably isn't money well spent.

» **Other board types:** An increasing number of man-made wood substitutes have made their way to market in recent years. Called *composite woods,* these engineered products (whose brand names include Trex, Veranda, and GeoDeck) are made up of mixtures of resins, regular lumber scrap or sawdust, natural plant material and fibers, and even recycled plastics. From a building perspective, working with them is just like working with real wood. These boards need little to no maintenance and come in a variety of colors. The downsides? This stuff is often prohibitively expensive and can be absurdly heavy. And because it's primarily a decking material, composites often aren't available in regular lumber sizes.

Knowing what to look for in a board

Certainly you've seen flannel–wearing contractors standing in the lumber aisle picking out their boards. The process usually involves taking a single board out of a pile, resting the far end on the ground, and holding the near end up to eye level. Then Joe Contractor squints one eye closed and stares down the length of the board, tilting it this way and that, looking for some magically secret clue signifying that this particular 2x4 has been uniquely crafted above all others to be a wall stud.

REMEMBER

If you hand–select individual pieces of lumber from the store (as opposed to ordering a bulk shipment where someone else pulls the materials to fill your order), it will be your job to weed through the boards you don't want in order to find the best ones for building. You don't have to be the Lumber Whisperer to pick out good wood; just look for a few obvious things:

» **Knots:** Knots in a piece of lumber are not inherently a problem. But steer clear of a piece of lumber with a knot on or at either end; it may not be as structurally sound as you'd like. Of course, if the board won't be used as a full-length piece, you can always cut that end off: Problem solved. Other knots you want to avoid are the big ones where wood is actually missing from the center. These may indicate a structural weakness that goes deep into the board.

THE PEOPLE VERSUS PRESSURE-TREATED LUMBER

There has been a great deal of passionate debate about the use and overall safety of pressure-treated wood. It is, after all, coated with and soaked in some heavy-duty chemicals, convincing some folks that it simply cannot be good.

Up until 2003, most pressure-treating of lumber was done with chromated copper arsenate, or CCA. Although the effects of long-term exposure to normal amounts of CCA were hotly contested, the use of the toxic preservative was discontinued. Most lumberyards now sell wood treated with either alkaline copper quaternary (ACQ) or copper azole (CA). Another type is treated with borates, engineered to resist termites. How these new chemicals compare to CCA in the long run has yet to be definitively determined.

So should you use pressure-treated lumber on your chicken coop? The debate lingers. There are those who argue that it's okay to use for posts that are partially buried in the ground, but never in the coop itself. Others say it's fine to use anywhere dampness or moisture might be an issue. Still others maintain that it should not be used anywhere at any time.

Unfortunately, a universally-correct blanket answer does not exist. Chicken-owners looking to practice true organic chickening will likely want to stay away from the chemicals in any pressure-treated lumber. Most longtime owners who have used pressure-treated lumber in and around their coops, however, have reported no health concerns that can be linked with its use. As a rule, chickens are not interested in pecking away at a piece of lumber, so their direct ingestion of any pressure-treating chemical would seem, at this point in time, to be a nonissue.

>> **Splits:** Some pieces of lumber will show splits or cracks at the ends. These indicate a weakened board and should be avoided, unless you know you'll be cutting that end off. The entire board isn't weakened — just the part that's actually split.

>> **Twisting:** While the milling process is supposed to crank out straight, uniform boards of any desired length, some boards have some warping or twisting to them. This is what Eagle-Eye Joe is looking for when he's squinting his way through every board in the lumberyard. Severe warping and twisting will be evident and could be tough to work with. A slight, gentle curve usually isn't a big deal. If a board seems pretty straight, it probably is.

>> **Wane:** *Wane* is a fancy-pants term for an edge of a board where the tree's bark used to be and may still be visible, even after milling. If that edge of that board will be exposed (and if it matters to you), leave that board for someone else. But as long as enough of the lumber's edge is intact for a nail or screw to bite into, there's nothing wrong with a little bark.

Shopping for Sheet Goods

Much of the wood you'll buy for your coop comes not in board form, but rather in large, thin sheets. These sheet goods cover big expanses quickly and cleanly, often serving as the cladding over a lumber framework, as on a subfloor, a roof, or walls. (For more details on how to work with sheet goods in these various applications, refer to Chapters 7 and 8.) In the following sections, we introduce the basics of sheet-good sizing, describe two popular types of sheet goods, and advise you to steer clear of a few other types.

Sizing sheet goods

TIP

Most sheet goods are sold in panels that measure 4 x 8 feet. (Some may also be found in smaller 2-x-4-foot pieces.) Full 4-x-8-foot sheets can be a real bear for the average DIYer to handle on the jobsite, and even in the parking lot of the lumberyard. If you won't need full 4-x-8 sheets for your coop (depending on the design you're using), consider asking the lumberyard to make some strategic cuts for you. Many places will make one free cut on a full sheet, under the auspices of allowing you to get it in your car more easily. So if you know you'll need your piece of plywood to be 4 feet by exactly 70½ inches, for example, ask your sales associate to make the free cut right at 70½ inches. The huge vertical panel saw in the store can do some of your work for you — and will probably make a much straighter cut than you ever could at home.

WARNING

Make sure that the 4-x-8 sheet you're buying is really 4 x 8 feet. Sheets marked "Sized for Spacing" actually measure slightly less than a full 48 x 96 inches. This is to assist in sheet-to-sheet assembly, where an expansion gap must be left between panels, as discussed in more detail in Chapters 7 and 8. But it can really screw you up if you don't know ahead of time that your panel is ½-inch short, so always measure ahead of time.

Checking out plywood and OSB

Here's a quick rundown of two commonly-used sheet goods (see Figure 4-1) to consider for your coop project:

>> **Plywood:** *Plywood* has become a bit of a catch-all phrase used to describe any type of wood product sold in a flat sheet. But true plywood is made up of several super-thin layers of laminated wood veneer, with new layers added (usually in odd numbers, to prevent warping) during manufacture until the desired thickness is reached (see Figure 4-1a). Each layer is oriented at a 90-degree angle to the layers above and below it, giving plywood incredible strength.

REMEMBER

You'll often see references to plywood thicknesses of ¼ inch, ½ inch, and ¾ inch. But these nice, neat sizes don't actually exist. Once you get to the lumberyard, you'll see that plywood is typically stamped with its *exact* thickness. So what you *call* ¾-inch plywood will *in actuality* measure $^{23}/_{32}$ inch. The former is just easier to say, so don't waste your time searching high and low for plywood that actually measures ¾ inch — unless you're talking about hardwood veneer plywood, which is often sized at full thickness, meaning that this ¾-inch plywood really *is* ¾ inch. (Confused yet?) Bottom line, to avoid problems, always personally measure any piece of wood before you buy it.

Construction plywood is made from one of several species of softwood, and is available in different grades for various applications:

● A grade of A or B denotes high-quality plywood meant for visible surfaces. These sheets have often been sanded smooth and contain few imperfections.

- C- or D-grade plywood is usually unsanded and more utilitarian in quality, meant for general construction applications where it won't be seen in the final product.

  Most of the time, a sheet of plywood will have two letter grades: one for each side. On a piece of A-C plywood, the A-side is nicely finished and suitable for an exposed surface, while the C-grade side is meant to be hidden. You'll also find B-C plywood or even C-D plywood (also called CDX) for purely structural purposes where neither side will be seen.

REMEMBER

Plywood meant for interior use won't last long in the great outdoors. A few good rains or prolonged exposure to high humidity will peel the layers of interior plywood like an onion. Exterior plywood uses water-resistant glue to adhere its layers together and resist rot.

>> **OSB:** Short for *oriented strand board,* this reconstituted product is made of large flakes (or "strands") of wood. Like plywood, thin layers are built up and glued together at 90-degree angles to create thickness and provide strength (see Figure 4-1b). OSB's low cost when compared to plywood makes it a popular choice as an underlayment or sheathing, where it won't be the finished visible surface. OSB, sometimes also called waferboard, is not considered a good all-weather option for exterior exposure.

REMEMBER

If you're planning to clad the exterior of your coop in some sort of siding, you can afford to be a lot less concerned about whether the sheathing is exterior–grade. Siding (as dealt with later in this chapter) requires some sort of weather barrier between it and the sheathing, so regular plywood or even OSB can be used to cut overall costs.

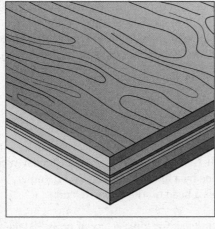 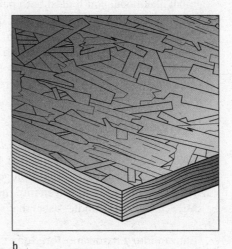

FIGURE 4-1:
Popular sheet goods for coop construction include plywood and OSB. a   b

Steering clear of certain sheet goods

WARNING

The following sheet goods are widely available but not recommended for coop construction (buy at your own risk!):

>> **Particle board:** Mix small wood shavings and sawdust with adhesives and resins, then press it all together in a sheet, and you have particle board. If you've ever bought a piece of ready-to-assemble furniture from a department store, you've probably seen it. It isn't all that strong (as evidenced by that bargain-bin bookcase you bought in college that was

sagging even before midterms) and tends not to hold nails or screws very well. Although it's cheap and plentiful, it swells dramatically and breaks down rapidly when exposed to water. It should probably be avoided altogether in coop construction.

>> **MDF:** *Medium-density fiberboard* is a waste-wood product composed of superfine wood fibers held together by binding agents. But this mixture is cooked and pressure-steamed to create a very dense material that's actually much stronger than particle board, OSB, and sometimes even plywood! It's very smooth, is knot- and blemish-free, and takes fasteners well, making it an exceedingly popular choice for many carpentry, furniture, and craft projects. But it's also quite heavy, is not at all water-resistant, and dulls saw blades and drill bits faster than other products. Plus, it's expensive, making it hard to recommend MDF as a practical coop-building material.

Getting Attached to Fasteners

Carefully selecting only the finest 2x4s and seeking out sheets of top-grade plywood is an excellent start to your project. But those pieces of lumber aren't going to just hold themselves to each other in the shape of a coop. You need to fasten them together, and that means nails and/or screws.

In all likelihood, you'll need both nails and screws for various steps of your coop build. While the nails-versus-screws argument often boils down to personal preference (especially on a smaller project like a chicken coop), you'll find times and applications where a nail simply won't do a screw's job and vice versa. It's unrealistic to think that you'll be able to pick up only one big container of just one type of fastener and build your entire coop with it. So here's a nuts-and-bolts look at nails and screws.

Notes about nails

Nails have been around for almost as long as humans have been working with wood. And at first glance, they haven't changed all that much. They're still slender pieces of metal, pointed at one end and driven into the wood by a blunt tool (usually a hammer, but not always, Mr. Screwdriver-Handle-User) that strikes a head on the opposite end.

But run into a hardware store sometime and ask for "just a box of nails." Make sure you do it when you have a few hours to kill, though, because you'll get peppered with follow-up questions: What kind of material are you nailing? What's the thickness of the material? How strong does it need to be? Is it indoors or out? Will this be a permanent or temporary attachment?

When it comes to nails, the variety is almost overwhelming. There are dozens of nail types of all different lengths and gauges, nails to be used only in very specific kinds of materials, specialized nails with chemical coatings, barbed ridges, or corkscrew spirals, even two-headed nails that look like mistakes that slipped by the nail factory's quality-control guy! In the following sections, we describe the range of nail sizes available and a few of the basic nail types you may want to use for your coop build.

REMEMBER

Nails are generally cheaper than screws of comparable size, and they tend to be less brittle and breakable than screws. Most pros find nails to be faster to drive than screws too — although many weekend warriors with sore arms, smashed thumbs, and rookie hammering skills may disagree.

Nail sizes

When you talk about a nail's size, you're really referring to two different things: the nail's length and the nail's thickness. A nail can vary in length from 1 to 6 inches. (Anything shorter than 1 inch is more properly called a *brad* or a *tack*; anything longer than 6 inches is technically a *spike*.) Talking about the thickness of a nail's *shank*, the long body that encompasses everything below the head, is less cut-and-dried. Nails typically aren't discussed in terms of their thickness alone; thickness and length go hand-in-hand. The longer a nail is, the thicker it is, too. If you want to upgrade in thickness, you have to bump up the length as well.

When selecting a nail for fastening two pieces of wood together (as you do in framing), the rule of thumb is this: Use a nail that goes all the way through the top piece of lumber and penetrates into the second piece (the *receiving* piece) by at least 1 inch. Going in more than 1 inch is fine (and holds better), as long as the nail doesn't fully penetrate the receiving piece and come out the other side. For example, if you're nailing a piece of ¾-inch plywood onto the flat side of a 2x4, which is 1½ inches thick, for a total of 2¼ inches, use a nail that's at least 1¾ inches long, for a minimum of 1-inch penetration. You could also use a 2-inch nail for a little extra holding power. A nail longer than 2¼ inches, however, would have no additional value, because it would come out the back of the 2x4.

WARNING

Don't get carried away with choosing a longer nail for that alluring-sounding "extra holding power." There's a tradeoff here: Remember that a longer nail is also thicker. Longer, thicker nails are harder to hammer in flush and wear you out faster over the course of a building day. And the thicker a nail is, the greater the chance you have of splitting the wood you're nailing into. For best results, stick with (or pretty darn close to) the rule of thumb above: Penetrate the receiving piece by 1 inch. If you find yourself splitting wood during a building project, try dropping down a nail size.

Nails are measured and packaged by their designated *penny* weight. See more on this often-confounding classification system in the sidebar "Why doesn't a 10-penny nail cost 10 pennies?" What you need to know about penny weight is this: The smaller the penny size, the shorter and thinner the nail is. At your local hardware store, you'll run into nails that range from 2-penny, abbreviated *2d* (which are 1-inch long), all the way up to 60-penny, or *60d* nails (which measure 6 inches in length). The good news is that most boxes of nails feature both the penny weight and length in inches right in plain English on the label, so you don't have to wear some secret carpenter's decoder ring when you go to the building supply center.

So which nail size is right for you and your coop build? There's no single right or wrong answer. But all things considered, the two most popular nail sizes are 8d and 16d, or 2½ inches and 3½ inches respectively. You could probably frame just about anything with these two nail sizes alone. (You'll notice, however, that in our coop plans in Part 3 of this book, we drop down one nail size and use 7d and 12d nails. A chicken coop's framing doesn't need to be quite as sturdy as your own home's, and you'll save a few cents on these just-slightly-smaller and easier-to-drive nail sizes.)

Nail types

There are more, but here are the nail types you typically might use to build a coop (check out Figure 4-2 for a visual reference):

>> **Box nail:** A box nail (see Figure 4-2a) features a thin shank and is therefore best when used in thin pieces of wood that may split easily, like fence boards or trim-work. To further aid in preventing splits, the pointed ends are usually blunted slightly. You'll find them in sizes ranging from 2d to 40d.

>> **Cement-coated nail:** A cement-coated nail (see Figure 4-2b) has been dressed with a thin layer of a heat-activated resin. As the nail is driven into wood, the friction creates enough heat to activate the resin and create a bond between the nail and the wood. It provides extra holding power, but is therefore harder to pull out if a mistake is made. If you like the added insurance that offers, spend a few extra pennies and go for cement-coated nails. Sizes typically range from 2d to 16d.

>> **Common nail:** A common nail (see Figure 4-2c) is, well, just that. It's a good general-purpose nail that works well for medium- to heavy-duty framing. It's pretty thick and ranges in size from 2d all the way up to 60d.

>> **Galvanized nail:** For use on the outside of the coop or in places that will otherwise be exposed to the elements, get galvanized nails (see Figure 4-2d). These fasteners, available from 2d up to 60d, have been coated with zinc to resist rust.

WARNING

If you want to use galvanized nails in pressure-treated lumber, be sure to read the fine print to see how the anti-rust formula was applied to the nails. *Electroplated* galvanized nails should not be used in pressure-treated lumber, because this method of galvanization results in a lighter coating of zinc that can be eaten away by chemicals in the lumber. *Hot-dipped* galvanized nails feature a heavier layer of zinc that allows them to stand up to the pressure-treating chemicals.

>> **Ring-shank nail:** A ring-shank nail is easily identified by a series of ridges that runs up and down the shank of the nail (see Figure 4-2e). Under magnification, these rings are seen to have a sharp lip that curls up toward the nail's head. (Think of an upside-down Frisbee.) These rings act like barbs, giving the ring-shank nail 50 to 100 percent more holding power than its smooth-shank counterpart, making it extremely difficult for the nail to work itself loose. That makes it an excellent choice for minimizing squeaks in a subfloor. But it also makes it brutally difficult to remove, and will bring a lot of wood with it if you succeed. Ring-shank nails are most often sold in sizes up to 60d.

>> **Roofing nail:** With its flat, oversized head (and often ringed shanks), a roofing nail (see Figure 4-2f) is designed to secure shingles to plywood roofing sheathing while doing minimal damage to the surface of the shingle itself. Roofing nails are unique in that the penny weight classification isn't applied to them; they're sold by their actual length, in inches. You'll find them in sizes ranging from ¾ inch to 1¾ inches.

>> **Spiral nail:** A spiral nail, sometimes also called a *helix* nail, has a distinctively-twisted shank (see Figure 4-2g). This unique shape helps it corkscrew into the wood as it's hammered for extra holding power. Great for flooring, not so great for pulling out. Spiral nails come in sizes 3d to 60d, and even up to 90d!

>> **Stainless steel nail:** If you're using cedar or redwood and the final appearance matters to you (and it likely does or you wouldn't be using cedar or redwood), you'll want to spend a little extra and get stainless steel fasteners (see Figure 4-2h). The natural acids in these woods reacts with other nail types and causes unsightly streaking and bleeding at every single nailhead. Stainless steel nails (and screws), available in all common sizes, help to keep the wood looking as good as the day it went up.

Stuff about screws

REMEMBER

For all the sheer simplicity and time-tested appeal that nails provide, many builders and scores of DIYers prefer to use screws whenever possible on a building project. Their spiral, helical ridges provide more holding power than nails and can often help draw two surfaces together with a little extra tightening. They're far easier to remove than nails and can even be reused. And because most people employ battery-powered screw guns to drive them (see Chapter 3 for more about these tools), they require less manual labor to use.

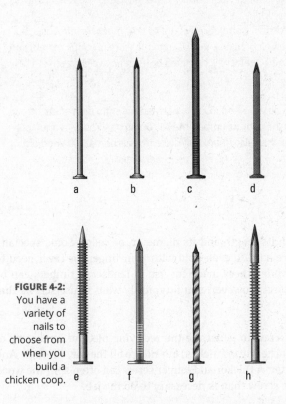

FIGURE 4-2: You have a variety of nails to choose from when you build a chicken coop.

You'll come across dozens upon dozens of different kinds of screws at your local hardware store. Screws tend to be even more specialized than nails. They're made out of a wider variety of materials and feature more unique finishes, thread orientations, overall lengths and thicknesses, and tool- and use-specific head shapes and configurations. In the following sections, we describe the most common screw sizes, types, heads, and drives out there.

WHY DOESN'T A 10-PENNY NAIL COST 10 PENNIES?

Nails are sold by weight and packaged together by length. But, as with so many other aspects of carpentry and woodworking, the nomenclature can be baffling. While you may be looking for a nail that's 3 inches long, what you'll end up buying is something called a *10-penny* nail. Even more confusing, it's written on the box as *10d*. What gives?

According to one version of the legend, the practice of using "pennies" to differentiate nail sizes dates back to medieval England, when the local blacksmith made nails by hand and sold them 100 at a time. One hundred 3-inch nails, for example, would set you back 10 pence, or pennies. The Roman coin called the *denarius* was the day's equivalent of a penny, and was commonly abbreviated to just its first letter, *d.* So a 3-inch nail, in time, became known as a 10-penny nail, or a 10d nail. A hundred smaller nails cost less and had a smaller number; 100 bigger nails cost more and had a larger number.

But conflicting research claims that the "penny" number refers not to how much the nail costs, but how much it weighs. The *pennyweight* is a unit of mass often used to measure precious metals, and is abbreviated *dwt.* Some say *that's* how a 10-penny nail, classified by its pennyweight, came to be written as 10d.

Whichever theory is historically accurate, all you need to know when a carpentry nerd starts throwing around the P-word is this: The larger the penny number, the larger the nail. Today, manufacturers still use the *d* in nail packaging, but they also (thankfully) list a nail's actual size in good old, easy-to-understand inches.

Screw sizes

Like a nail, a screw's size refers to both its length and its diameter, or *gauge*. Some specialty screws can be so teeny-tiny that they're hard to even hold with your fingertips (ever need to replace one in a pair of eyeglasses?), while others made for use in landscape timbers can be 10 inches long! For a coop-building project, however, you'll typically want to use screws that are between 2 and 3½ inches long.

As with nails, you typically want your screw to penetrate the receiving piece of lumber by at least an inch (some builders prefer 1¼ inches or 1½ inches) but not go all the way through. And again, like nails, bigger isn't always better: A thicker and longer screw can often split the wood you're driving it into, so don't use more screw than is necessary to do the job.

WARNING

Although you can use either nails or screws (or some of each) for your coop project, be careful when you substitute one for the other! A 3-inch screw can often be much larger in diameter than a 3-inch nail and end up splitting your lumber. When subbing screws for nails, be sure that you're matching diameter first, and then come as close as you can in length. For example, our plans in Part 3 call for lots of 7d nails, which measure 2½ inches in length. But if you prefer to use screws, we recommend using screws that are just 1¼ inches long. The diameter is likely to be much closer to that of a 7d nail, and the shorter screw will still provide the holding power needed to do the job.

Screw types

For building a chicken coop, you need to decide among just a few basic screw types, listed here in alphabetical order (see Figure 4-3):

>> **Deck screw:** The favored screw for any outdoor construction application, deck screws (see Figure 4-3a) are galvanized or otherwise coated specifically for exterior use. Their long length (sometimes over 3 inches) ensures that each screw goes through 1½-inch-thick material and sufficiently penetrates the second surface. A deck screw's tip is super-sharp to eliminate the need for drilled pilot holes and to improve ease of driving.

WARNING

Although they look quite similar to deck screws and are far less expensive, don't be tempted to substitute drywall screws during your coop build. Drywall screws are much weaker then deck screws and don't have nearly the same holding power for often-heavy pieces of lumber. Further, drywall screws aren't treated to withstand exposure to the elements, meaning they'll corrode and rust much faster.

>> **Lag screw:** Thicker than typical screws (between ¼ inch and ½ inch in diameter), lag screws (sometimes called lag *bolts*) are generally used for heavy-duty framing, like for attaching a rim joist to a vertical timber post (as described in Chapter 6). Lags (see Figure 4-3b) require a pilot hole, because they can easily split lumber, and they have hexagonal heads that must be driven with a wrench or socket. Stainless steel or hot-dipped galvanized lag screws work best for outdoor applications.

>> **Wood screw:** These screws (see Figure 4-3c) are made for securing thin materials to thicker pieces of wood. You can identify them by their threads, which reach only 75 percent of the way up the screw, leaving an unthreaded portion of shank just below the screw's head.

Wood screws can be found in over a dozen diameters, from a small and skinny #2 ($\frac{3}{32}$ inch in diameter and ¼ to ½ in length) to a big and beefy #14 (¼ inch in diameter and 1 to 2¾ inches in length). The most common sizes you might use on a coop build are #6, #8, and #10. They can be difficult to find in sizes over 2½ inches, though, so they're not recommended for framing applications. They're better for fastening wood trim or securing hardware like latches and hinges.

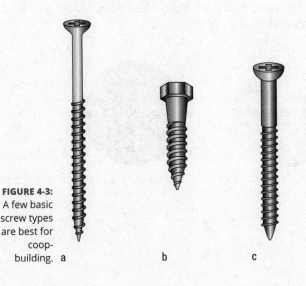

FIGURE 4-3: A few basic screw types are best for coop-building.

a   b   c

Screw heads and drives

TIP

You'll see several different head types on screws as you shop. Many screws have rounded, oval, or pan heads that will stick up out of the finished surface, even after being fully driven. Deck screws feature flat heads. These *countersunk* heads bury themselves in the wood as the screw is driven, leaving behind a perfectly flat surface. Lag screws have hex heads that need to be turned with a wrench or socket and will stick up out of the finished surface.

Finally, a word about the screw drive types you'll run across. *Drive* refers to the configuration of the head and the tool (or bit) that's required to actually turn the screw. Again, there are more than a dozen specialized drive types for various applications. The average DIYer, though, doesn't need to be concerned about the more exotic drives, and can focus on one of the basic three, as seen in Figure 4-4:

» **Phillips:** Named after the company that pioneered its use, the Phillips-head screw (see Figure 4-4a) features a cross shape on its head. The cross has a pointed tip in its recesses that makes it self-centering, so a Phillips screwdriver or bit will slip into place cleanly. But that pointed tip also makes it easy for the tool to pop out if you're trying to really tighten a Phillips screw down.

» **Slotted:** The basic slotted-head screw is recognized by its head's single, straight-line channel (see Figure 4-4b). It works with a flat-bladed screwdriver or bit. Most carpenters bypass slotted screws for use in construction projects, because a slotted bit tends to slip out of the screw when used with a screw gun.

» **Square:** Also called a Robertson head (after the guy who invented it), a square-headed screw (see Figure 4-4c) has a square hole. (But you saw that coming a mile away, right?) It's designed to maximize torque to the screw when turning and minimize the chances of the bit slipping out unintentionally. Square-drive heads and the bits that turn them are becoming more and more popular and easier to find in most hardware stores and home centers.

TIP

Each kind of screw head comes in different sizes. The slot on a slotted-head screw can be one of three common sizes: #6-8, #8-10, or #10-12. The same goes for Phillips and square-head screws, sized as #1, #2, and #3. The wrong size screwdriver or bit *may* turn the screws you have, but it won't do so easily, and you won't get good, tight connections. Always double-check to make sure your screws match your tool.

 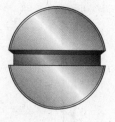 

FIGURE 4-4:
Common screw drive types.

a   b   c

Figuring Out Flooring

For the vast majority of coops, the flooring of the housing structure is constructed from basic plywood or some other sheet wood, as described earlier in this chapter. These materials are inexpensive and easy to work with, and therefore a sound choice. (Flip to Chapter 7 for instructions on building a subfloor using plywood.)

But the wise coop-owner is already thinking ahead, even at this early planning stage, about long-term use, care, and maintenance. That coop you're about to build will be sparkly, shiny, and new only until the moment you let a chicken inside. And that plywood floor you precisely measured, carefully cut, and meticulously fastened into place will quickly become, well, a toilet. So cleaning that floor will become a major concern. Even though you'll almost certainly want to use some type of loose bedding on the floor of your coop, many people like to make the floor itself as easy to clean as possible. Here are a few popular choices for flooring:

>> **Linoleum/vinyl flooring:** While a wooden sheet material is easy to work with, even exterior-grade plywood isn't meant to be repeatedly and thoroughly hosed down on a regular basis. A layer of cheap linoleum or similar flooring product on top of a solid subfloor (the kind described in Chapter 7) can work wonders in making cleanup a breeze. (But you'll still want to throw a layer of soft bedding material on top of it to keep it from becoming a chicken Slip 'n' Slide.)

TIP

Here's a great place to improvise with some vinyl flooring taken out of a recently-renovated bathroom, a rubber liner rescued from a neighbor's old pond, or a bargain from the close-out corner of your local big-box warehouse store. Your hens won't mind if their linoleum is "so last year" or even a holdout from before disco was king.

TIP

More than a few coop-owners have fallen in love with a product called Glasbord. A fiberglass-reinforced plastic that comes in panels, Glasbord won't mold, mildew, or rot. It's extraordinarily resistant to moisture, staining, and scratching. It's even lightly pebbled, providing a bit of traction under your flock's feet. You may need to do some searching to find a local distributor, but those who have used Glasbord on a coop floor swear by it.

>> **Concrete:** Many people choose to build their coop on an unused corner of a patio and not construct a floor at all. Concrete is a perfectly acceptable floor for a chicken coop. It's easy to clean and can be hosed down without worry. But unless you're pouring a concrete slab way out in the yard for this sole purpose, a concrete floor generally means your coop is located pretty close to the house, and that can be a deal-breaker for even the most enthusiastic chicken owner (or the most supportive chicken owner's spouse).

>> **Wire mesh:** Some very small coops have wire floors. (You can find more detail on wire products later in this chapter.) From a maintenance standpoint, this can be a godsend, as chicken waste falls through the wire mesh and disappears below. But remember that the waste doesn't vanish; it still goes somewhere, and that "somewhere" is usually just under the coop. So now *that* will need to be cleaned up from time to time. Some coops employ some sort of a removable tray under a wire floor, whereby coop cleanup involves pulling the tray out, cleaning it, and replacing it without ever interrupting the chickens' busy schedules.

Most who choose to employ wire flooring use fine wire mesh in place of a solid subfloor. If this idea appeals to you, be sure to use a wire mesh with very small holes. This allows you to top the wire floor with loose bedding material that won't fall through the mesh as easily. And smaller holes in a wire floor minimize the chances of a bird getting her toenails caught, which can be very painful.

>> **Dirt:** Plenty of coops are built directly on the ground, using good old Mother Nature as the floor of the shelter. A dirt floor may sound like the ultimate in low-maintenance, but it has its downsides, too. Expect an increase in pests, and possibly even a burrowing predator or two. A dirt-floored coop can also be a nightmare to really clean out well (and there will be times you'll want to do just that).

REMEMBER

Whatever material the floor itself is made out of, practically all coop-owners choose to cover it with some type of loose bedding material. It's perhaps better thought of as *litter*: something that makes the floor of the shelter easier to clean and is topped off or replaced on a regular basis. The ideal bedding is absorbent and acts as an odor combatant (but only to a certain extent!). Most chicken-owners use wood shavings (pine being the most commonly available and the cheapest), but others have had success with sawdust, horse bedding, sand, and even finely-shredded landscape mulch.

Wondering About Walls

The walls of your chicken coop offer perhaps the greatest amount of flexibility in terms of what material you elect to use. For many utilitarian coops, simple plywood (as discussed earlier in this chapter) will suffice. Other coop-builders may choose something else, either to make the coop look more attractive or in order to utilize a material they happen to already have on-hand. Check out this short list of some of the more widely-used wall materials (and see Chapter 8 for details on wall construction):

>> **Siding:** Most commonly made of wood, vinyl, or fiber cement (brands like HardiePlank), siding is easy to install and quite versatile. You typically install horizontal planks over plywood sheathing and a weatherproofing layer of felt paper or housewrap. Using siding can allow you to match your own house or customize your coop with color.

>> **Exterior paneling:** For a rustic look, consider a popular plywood product often referred to by contractors as T1-11 (pronounced "tee one-eleven"). With vertical grooves every 8 inches or so, this exterior-grade plywood is specifically designed for projects where the final look matters. It's sold in 4-x-8-foot sheets, just like regular plywood, with tongue-and-groove edges to allow sheets to fit together cleanly with no visible seam. It's usually unfinished, so whether you paint or stain it after the build is up to you and the chickens.

Getting to the Root of Roofing

Putting a roof over your flock's heads is practically the whole point of building a chicken coop. But surprisingly (or perhaps not surprisingly, if you stop to think about it), there simply aren't that many materials you'd want to use to roof a coop. Here's the rundown:

>> **Shingles:** Most coops are roofed just like a home, using a multi-layered approach (see Figure 4-5) to make the structure weather-tight.

- The overlapping shingles, the outermost layer that's exposed to the elements, are most commonly made of asphalt, but can be found as cedar, metal, or even slate.

- Underneath the shingles is a layer of roofing paper, also called *felt* or *tar paper*. This underlayment serves the critical purpose of acting as a moisture barrier for the shingles. Shingles can become drenched during heavy rains; roofing paper keeps that water out of your chicken coop.

- The bottom layer of a typical roof is plywood sheathing, the solid surface that provides rigidity to the structure.

Shingles are fairly inexpensive and easy to install, requiring no special tools or hardware other than specialized roofing nails (see the earlier section "Notes about nails").

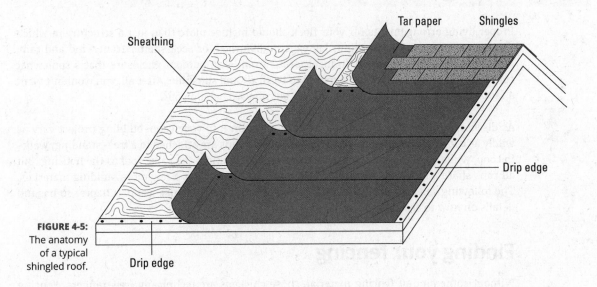

FIGURE 4-5: The anatomy of a typical shingled roof.

Labels: Tar paper, Shingles, Sheathing, Drip edge, Drip edge

>> **Corrugated panels:** Corrugated panels (as shown in Figure 4-6) are an inexpensive and lightweight alternative to shingles for many outdoor structures like patios, carports, greenhouses, toolsheds, and — drum roll, please — chicken coops! These sheets, usually made of metal or translucent fiberglass, come in a variety of sizes and colors.

For those who like corrugated roof panels, there are upsides. The translucent panels let in diffused light, which can aid in your chores and coop maintenance. They eliminate the need for working with asphalt shingles, which many people simply dread. And there's no sound quite like a steady rain hitting the top of a panel roof. (But are you really going to be hanging out inside your chicken coop during a summer storm just to soak up the sounds?)

On the negative side of the ledger, corrugated panels can be tough to work with. On a small-scale structure like a coop, you'll almost certainly have to cut the panels to size, which can be a tricky proposition for a DIYer because they're made of either metal or brittle fiberglass. Roof panels have a reputation for being easy to break, a real consideration if your coop will be located under tree limbs. And finally, you'll have to buy installation accessories. To offer proper weatherproof protection, fiberglass panels must be used with special support pieces called *closure strips* and all panels require specialized fasteners that typically include some sort of rubber grommet to help create a watertight seal at each screw location.

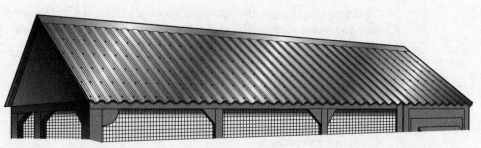

FIGURE 4-6:
Corrugated roofing panels are lightweight and economical, but do require special hardware.

Researching Your Run

Proper living arrangements for your flock should include more than just a structure in which they can sleep protected at night, lay eggs undisturbed, or seek shelter from wind and rain. Most coop-builders also want to incorporate a *run* — an outdoor enclosure that's somewhat confined and protected, yet still exposed to fresh air and sunshine. After all, you wouldn't want a fantastic new home with cozy bedrooms but no yard, would you?

As discussed in Chapter 2, ways to incorporate a run into your coop-building project vary as wildly as coop designs themselves. Some expansive runs are attached to a free-standing, walk-in coop, while an all-in-one coop includes a small run under the same roof as the housing. But all runs share a few basic features that require a closer look at a few more building materials. The following sections cover the basics of what you'll need to raise a run. (Chapter 10 has full details on how to assemble a run.)

Finding your fencing

Without some kind of fencing material, those chickens are technically free-rangers. Fencing protects your chickens from predators, while letting them feel like they're roaming unencumbered in the great outdoors. You've got quite a few options when it comes to your fencing material, and some are quite a bit better than others.

Making a smart selection about fencing material usually boils down to two factors: the material's strength and the size of the openings. When considering strength, remember that your fencing has to do more than just keep your flock contained. Even more important, it must keep predators from prying their way in, so stronger is better. Second, look at the openings in the fencing material you're considering. Just because a raccoon can't squeeze all the way through some types of fencing doesn't mean it offers adequate protection for your birds. A predator doesn't have to get all the way into a run to wreak bloody havoc. Nimble predators love to reach in with their arms and wave their razor-sharp claws around, trying to slash at chickens who wander too close. We especially like products with narrow 1-x-2-inch openings or smaller to keep four-legged Freddy Kruegers at bay.

Here are facts you should know about fencing as you design your run.

>> **Chicken wire:** It seems like a no-brainer that a product actually called *chicken wire* should be the most obvious and best choice for fencing. Yes, it's easy to find, it's inexpensive, and

it's quite malleable and therefore a cinch to work with. But most seasoned chicken-owners (meaning "veteran" owners of chickens, not owners of chickens that have been sprinkled with salt and pepper) will tell you that true chicken wire, with its thin-gauge wire and wide hexagonal openings (see Figure 4-7a), is far too lightweight to fend off any serious attack from a would-be chicken-snatcher.

>> **Nylon/plastic:** Nylon and plastic are even more flexible than chicken wire, making them easier on your hands during building. Your options here may be spread all over your home center, from light garden netting all the way up to the heavy-duty orange snow fencing you often see on a highway construction project. Like chicken wire, though, plastic and nylon fencing tends to be great for keeping your chickens in — but not so great for keeping predators out.

TIP

If your neck of the woods is home to birds of prey, you may find garden netting or a similar lightweight material to be a low-cost and not-terribly-unattractive canopy for covering the top of the run, even if the walls of the run are made of something sturdier. Hawks and owls should be adequately deterred, but this type of flimsy material won't offer much defense against raccoons and other varmints that can climb the fence walls and go right through a nylon canopy.

>> **Hardware cloth:** Often stocked right next to the chicken wire at your hardware store is something called *hardware cloth* (see Figure 4-7b). Wire stock — usually heavier than that of chicken wire — is woven and welded into a grid (with normally either ¼-inch- or ½-inch-square openings in the mesh), and it's often a sturdier choice for fencing a chicken run.

>> **Welded wire:** Welded wire looks a lot like hardware cloth, but it's usually made with even heavier wire. While it's stronger than hardware cloth, its openings are generally much larger as well, leaving your chickens more susceptible to the long and nimble arms of a raccoon.

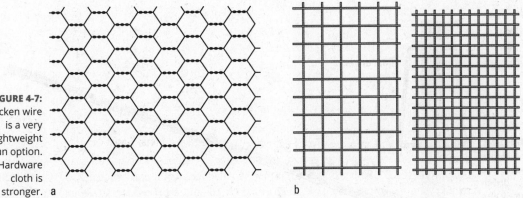

FIGURE 4-7: Chicken wire is a very lightweight run option. Hardware cloth is stronger. a

b

TIP

Many coop-owners like to double up, layering a 1- or 2-foot-tall run of hardware cloth along the bottom of a full fence of welded wire. This two-ply approach takes advantage of the considerable strength of welded wire for the overall structure, but keeps precious chicks on the ground more safely contained behind the hardware cloth's tighter mesh.

>> **Fencing panels:** Chicken wire, nylon and plastic, hardware cloth, and welded wire all come in rolls and are cut to the length needed. Fencing panels, on the other hand, are rigid pieces of fencing material sold in uniformly-sized sections. You'll find panels made of wood, wire, or even plastic or vinyl. They typically provide more strength than any rolled material, but they do require more fence posts, because each side of each panel needs to be properly supported.

Picking your posts

Whatever fencing material you select, it will need to be supported — in most cases, fastened to vertical members. On a small all-in-one or tractor coop (see Chapter 2), these vertical support posts may be 2x4s that help comprise the coop. For a larger run, you're probably looking at real fence posts. If that's the case, you basically have two options:

>> **Wood:** Want to get in touch with your inner cowboy? Spend a day on your back 40, sinking a long line of wooden fence posts. These can be regular 4x4s or the more ranch-appropriate rounded corral poles. Either way, plan on a long day digging post holes, a long day mixing and pouring concrete, and a long day setting and bracing the posts to properly anchor them in the ground and provide your run the support it needs. (Unfamiliar with these tasks? Flip to Chapter 6 for details on what to do.)

>> **Metal:** The classic metal fence post, often called a *T-post,* (see Figure 4-8) is a piece of steel, typically painted green, with a spade-like plate that helps stabilize the post in the earth. A series of tabs running the length of the post holds the wire fencing material in place. This kind of post can be driven into the ground with a heavy hammer or a specially-designed tool called a post driver, also shown in Figure 4-8.

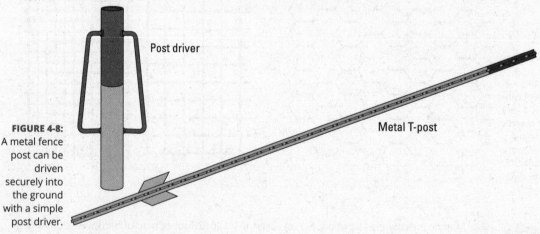

Post driver

Metal T-post

FIGURE 4-8: A metal fence post can be driven securely into the ground with a simple post driver.

Adding It Up: Estimating the Amount of Materials You Need

Once you decide on *what* materials to use on your coop build, you need to figure out *how much* of each material you'll need. If you're planning to construct one of the coops we've laid out in Part 3 of this book, it's as easy as turning to your particular coop's materials list, where we break out exact quantities of everything.

Flying solo on your own coop design? If you've sketched out your coop design, estimating your materials should be pretty easy. Working off some sort of drawing or sketch is a practice we highly recommend. It doesn't have to be a precise, to-scale work of blueprint-quality art; it just serves as a basic roadmap for you to follow as you build.

Break down your coop into a list that features each type of lumber you'll need to build it: 5/8-inch OSB, 2x4s, 1x3 trim, and so on. Now just add up the total measurements of each lumber type, keeping in mind the standard sizes of each material you'll encounter at the lumberyard: 4-x-8-foot sheets for OSB and plywood, 8-foot lengths of board lumber, and so on. For instance, if your coop design requires a total of ten 4-foot-long 2x4s, you'll want to buy five 8-foot lengths of 2x4; cutting them each in half will give you the ten you need.

REMEMBER

How important is it to know how cuts of lumber are sold at the building supply center? Consider this example: Say your coop design calls for 16 3-foot-long 2x4s. That's a total of 48 feet, so you stroll off to the lumberyard and buy six 8-foot lengths. "Piece of cake," you say smugly, until you run short during cutting. Here's why: Each 8-foot length will yield two of your needed 3-foot pieces, along with a 2-foot length of scrap. Cutting all six lengths that you bought will give you just 12 of the 16 pieces required. You'll now have to go back to the store to buy two more 8-footers. Estimating quantities isn't just a matter of adding up the total *linear feet,* it's knowing how much usable lumber you'll get out of each board you buy.

Once you arrive at your totals, all builders will tell you this golden rule: "Add 10 percent." Even if you've been meticulous in calculating everything down to the exact inch, it's always advisable to buy slightly more than you think you need. This overage covers a multitude of unforeseen surprises: a bad cut that wastes a piece of wood, a defective board you didn't notice in the lumber aisle and don't want to use after all, extra scraps needed here or there, even simple mathematical errors.

If your sketch shows that you need 20 8-foot lengths of 2x4, buy 22 of them, just in case. If you don't use those extra two, you can always return them to the store at your leisure for a refund (or keep them handy for a future project). If you run out in the middle of the build because you cut your estimate too close, though, you just bring the whole project to a grinding halt.

Estimating something like nails or screws can be trickier; it's hard to pinpoint how many you'll really need. Overestimate in order to factor in eventualities like bent nails and stripped screws. Most fasteners are sold in boxes of 1 or 5 pounds. Need "just a few" nails for the trim-work? Go with a 1-pound box. Framing an entire coop, though, will likely eat up a 5-pound package pretty quickly. And usually, by the time you buy two or three 1-pound boxes, you've paid for a 5-pound box. Having leftover fasteners stocks your workshop for the next project; coming up shy in mid-build is just poor planning.

Chapter **5**

Building Your Carpentry Skills

Builders have a saying: "Don't blame the tool; blame the carpenter." It's an oft-repeated reminder that if the project doesn't turn out quite right, it's more likely due to shoddy workmanship or simple carelessness than the wrong size hammer or a dull saw blade. (A thorough craftsman would have selected the appropriate hammer and checked the sharpness of that blade before starting!)

You can have a huge workshop stocked with the newest, most accessory-laden, best-built tools that money can buy. You can do exhaustive research on building materials, carefully hand-selecting the finest A-grade boards and industrial-strength fasteners. But those things don't guarantee a quality coop. In fact, they don't mean much at all if you don't know what you're doing.

This chapter offers tips and tricks for effectively and safely using some of the most common carpentry tools used to build a chicken coop; check out Chapter 3 for an introduction to these tools.

REMEMBER

The aim of this chapter is to give you confidence when it comes to using tools to build your coop. Sometimes, although the spirit may be willing, the flesh may be weak, especially when faced with a tricky saw cut. At any point during your build, if you feel like a particular step or technique simply exceeds your abilities or comfort level, don't hesitate to call for backup from a contractor, handyman, or well-practiced friend or neighbor.

Measuring and Marking Materials

Using a tape measure and pencil may seem as straightforward as it can possibly get. But professional builders have a few secrets — for even so rudimentary a task — that help them get precise results. Here are some ways to ensure that you're right on the mark every single time.

Reading the tale of your tape

Here's a make-or-break moment for any DIYer: How well can you read your tape measure? This is perhaps the single most vital skill for building a chicken coop, or anything else, for that matter. All tape measures show inches and the basic quarter-inch fractions thereof. Many tapes break inches down even further into eighths and sixteenths. Some tapes even show you thirty-seconds of an inch!

No matter what increments your tape measure uses, take the time to find out how to accurately read it before you start measuring and cutting. "Ten inches and eleven-sixteenths" is foolproof. "Ten inches and three little lines past the big line halfway between ten and eleven" is a mistake waiting to happen.

REMEMBER

"Measure twice, cut once" is a cliché for a reason: It's darn good advice. Manufacturers have yet to invent a tool that corrects a mistakenly-cut piece of lumber. Once you cut, it's irreversible. So double-check every measurement for accuracy before you fire up that saw.

"V" marks the spot, and "X" marks the trash

After they've measured precisely, most pros mark their lumber with a "V," placing the point exactly on the spot to be cut (see Figure 5-1). A simple slash mark may veer one way or the other, causing confusion when it's time to make the cut. A "V" points to an unambiguous spot for the blade to hit.

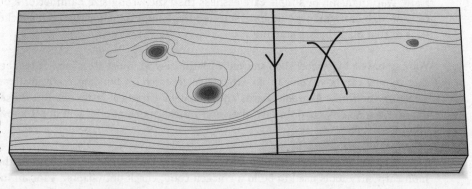

FIGURE 5-1:
Marking a board with a "V" and an "X" makes precise cutting as easy as A-B-C.

TIP

While a small "V" gives you a point to aim for with your blade, most DIYers prefer to cut along a straight line. We like to use a straightedge at the point of the "V" and, with our pencil, extend the line across the entire width of the board. Using a tape measure as a marking straightedge can be tricky, because the pencil will often skip right over the thin metal of the tape. The solution? A speed square. This versatile tool makes a superb straightedge. Hold the lipped edge of the square against the top or bottom of the board, line up the perpendicular edge with your "V," and scribe a long, easy-to-see line with your pencil. (We discuss squares in more detail later in this chapter.)

Say you have a 48-inch board, and you need a piece that's 24½ inches long. You measure it out, mark the spot with a "V," and make your cut. As you finish the cut, both pieces of wood fall to the ground. So you pick up the 24½-inch piece and get ready to nail it in place. Or is that the other end of the board, the 23½-inch piece of scrap?

After they mark their cut spot with a "V," professional builders often scribble an "X" onto the side of the board that they don't need (as seen in Figure 5-1). This shows them at a glance which end of the board is the scrap wood.

Chalk it up

A chalk line conveniently transfers a straight line across a long distance, as when marking a sheet of plywood. But for someone who's never used one, it can be a baffling little device. Here (and in Figure 5-2) are some tips:

1. **Make your marks.**

   Make a small pencil line at either end of the piece to be marked.

2. **Shake it up.**

   Always shake the closed container (the *chalk box*) before pulling the string out. The chalk can get caked up in there; a good shake loosens it up and helps coat the string with fresh chalk.

TIP

   Chalk is sold in a few different colors, red and blue being the most common. Choose a color that will be easy to see on the piece you're marking.

3. **Hook the string over the workpiece.**

   Clip the metal hook over one end of the piece to be marked (as shown in Figure 5-2a). Most string lines feature an open hook, allowing you to center it precisely on your pencil mark.

4. **Pull and snap.**

   Stretch the line across the work-piece and pull at the other end, making sure the string lines up with your pencil mark and *holding the line taut* (as shown in Figure 5-2b). The line should be just touching the surface. With your other hand, pull the string straight up and let it snap against the piece, leaving a line of brightly-colored dust for you to refer to as needed (see Figure 5-2c).

5. **Wind the string up and repeat.**

   You may get a second snap out of a pulled chalk line, but after that, you'll have lost enough chalk that your lines will probably begin to get faded and hard to see. Wind up the string to recoil it inside the box, and start over with Step 1.

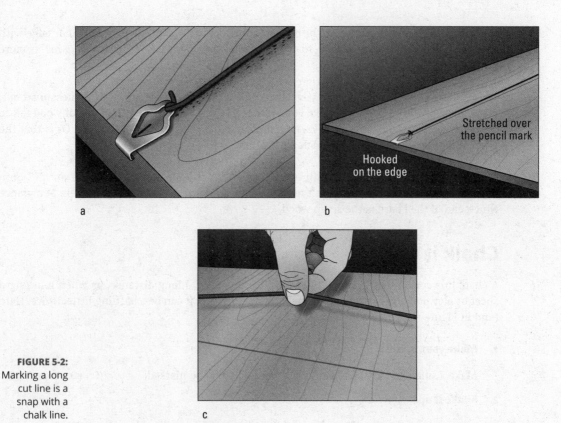

Stretched over
the pencil mark

Hooked
on the edge

a

b

c

FIGURE 5-2:
Marking a long
cut line is a
snap with a
chalk line.

Cutting Wood Safely

Working with saws is an essential part of any building project, but it's also dripping with danger. These machines are powerful, with carbide-tipped blades spinning at outrageous speeds, often within inches of various bodily appendages. They're loaded with safety features, but ultimately, your safety is up to you. If you ever feel uncomfortable about using a particular saw or performing a specific cut, stop! There's no shame in asking for help from someone more experienced, especially if it means keeping all your fingers.

REMEMBER

It's easy to become complacent after running a saw a few times, but all it takes is one careless moment to ruin everything. Stop and think about where your hands and fingers are before squeezing that saw's trigger. Be mindful of your stance, your balance, and your surroundings. Visualize the cut all the way through to the end, thinking about what might get in your way mid-cut: your sawhorses, your clamps, your straightedge, the power cord, your leg, and so on.

The following sections describe some more rules to follow when firing up a power saw.

Selecting saw blades

Not all saw blades are created equal. You need a blade that's engineered to slice through the type of material you're cutting.

Wood blades vary wildly. Many blades can cut through most types of wood, but using the appropriate blade results in smoother, easier, and more accurate cutting, with less wear on the saw's motor. And, it's safer.

TECHNICAL
STUFF

Special abrasive and diamond-tipped blades are even available for cutting through concrete, although you aren't likely to need one of those for a chicken-coop project. If you're building a chicken *bunker*, on the other hand. . . .

REMEMBER

You'll most frequently use a circular saw and/or a miter saw on a coop project (Chapter 3 notes some specialty saws you may want to consider). A miter saw is so powerful and so fast that few DIYers ever need to worry about changing the blade. The blade that comes on the miter saw is as "all-purpose" as you'll likely ever need. Circular saw owners, on the other hand, have to deal with dangerous issues like kickback and binding (see the following section) that can be caused by using the wrong blade type. You should not only know how to change your circular saw's blade, but be prepared to do so depending on what you're cutting.

Here's a who's who of circular saw wood blades, also seen in Figure 5-3:

>> **Crosscut:** To *crosscut* a board means to cut it across the wood grain, perpendicular to the visible pattern of lines in the wood itself. "Regular" cuts across the 3½-inch face of a 2x4 (or the 5½-inch face of a 2x6, and so forth) are crosscuts. A crosscut blade (as shown in Figure 5-3a) typically has 48 or more teeth for a smooth cut, and the top of each tooth is angled to slice through the wood more easily.

>> **Rip:** Identified by its larger, straight-topped, and less numerous teeth (as few as 24) and deeper *gullets* — those are the "dips" cut out of the blade itself in front of each tooth that allow waste material to be cleared from the spinning blade — a blade meant for ripping (as shown in Figure 5-3b) is used for making long cuts *with* the wood grain. With fewer teeth, a rip blade's cut is generally not as fine as a crosscut.

>> **Combination:** Exactly what it sounds like it would be, a combination blade (as shown in Figure 5-3c) takes the best attributes of a rip blade and a crosscut blade and puts them together. It often has teeth similar to those on a crosscut blade, separated by the deep gullets associated with a rip blade. If you want to try to use just one blade for the entire coop build, this is it.

>> **Plywood:** To tackle the thin, laminated sheets of plywood, blade manufacturers have designed special blades that feature many very fine teeth (sometimes 100 or more). The large number of small teeth on a plywood blade (as shown in Figure 5-3d) helps prevent splintering of the wood as it slices through.

TIP

If you're cutting a lot of plywood that will be seen, a dedicated plywood blade is worth the investment. For just a few quick cuts where a laser-like edge isn't critical (in subfloor decking, for instance), a crosscut or combination blade is likely fine. A rip blade, however, should never be used on plywood.

Cutting board lumber

Sawing through two-by-whatever lumber on the jobsite is usually done with a circular saw or a miter saw. While the tool does most of the work for you, there are techniques you can use to cut down on mistakes. (Get it? "Cut" down on mistakes? Fine. Groan now, but you'll use these tips later.)

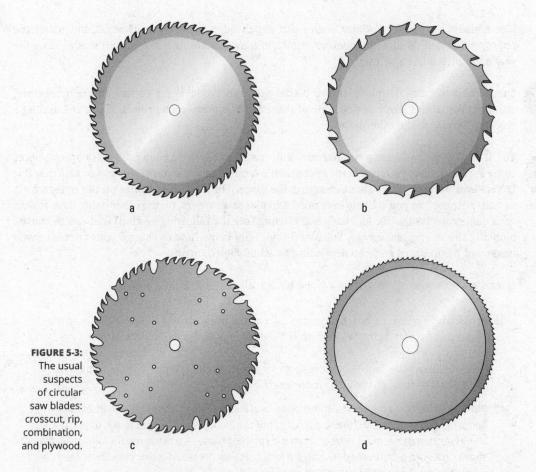

FIGURE 5-3:
The usual
suspects
of circular
saw blades:
crosscut, rip,
combination,
and plywood.

a   b

c   d

Using a circular saw

Keep this checklist in mind when using a handheld circular saw to cut board lumber:

1. **Adjust the depth.**

    You can manually move the blade to expose either the entire bottom half of the circular blade, or just a sliver. Open the blade guard (with the saw unplugged!) and gauge its depth, relative to the thickness of your wood. It should cut all the way through, but just barely. Expose more than ¼ inch of blade, and you increase the risk of *kickback*, a dangerous condition whereby the saw violently jerks backward in your hands. Exposing too much blade also tends to splinter the wood instead of leaving a nice, smooth cut.

    WARNING

    Watch where you're standing when operating a circular saw. Too many weekend warriors have been seriously injured because their saw kicked back — and right into them. Get in the habit of positioning your body so that it's not directly lined up with the blade. You should be reaching outside the width of your body to control the saw from behind.

2. **Stabilize the board.**

    You can't beat a pair of sawhorses for this task. Situate the board with your cut line hanging off the edge of one sawhorse, leaving the scrap end unsupported. This end will now fall safely during the cut.

TIP

If the piece you're cutting has a long overhang of more than a foot or so, a makeshift cutting table works best. Set up a sturdy piece of plywood across your sawhorses. Lay the board to be cut on top of the plywood. Find your cut mark and slide a scrap piece of lumber under your board on the longer side of your cut mark. This will allow your cut to open and spread away from the blade but only fall a half inch or so before being supported by your plywood table.

WARNING

What you don't want to do is make a cut between your only two sawhorses. This will cause the two halves to fall in on each other, where they can easily pinch and bind the blade, causing kickback.

3. **Start the blade early.**

After you plug in your saw, you're ready to cut. Make sure the blade isn't touching the wood at all when you squeeze the trigger. Allow it to run for a second or two to get up to full speed before beginning your cut. This is easier on the saw's motor and prevents splintering along the edge of the wood.

4. **Consider the kerf.**

In carpenter-speak, the *kerf* refers to the minute amount of wood removed by the width of the blade itself during a cut, usually $\frac{1}{16}$ to $\frac{1}{8}$ inch. A thick blade removes more wood than a thin blade, and you must take this into account when you line up the blade with your cut mark. Align the blade so that the "V" (or line) you so meticulously measured and marked is the true end of the board, even after the cut is complete. Mark an "X" on the scrap side of your line, and position your blade on the same side of the line as the "X." "Erasing" that line with the cut can leave your board a fraction of an inch shorter than you intended.

5. **Go slow and steady.**

WARNING

The blade does most of the work; you're merely guiding it. Note the use of the word *guide* instead of *push*. You need to exert some amount of pressure, but forcing the blade into the wood can result in the blade binding and coming to a frighteningly sudden halt, sometimes with the accompanying stench of a smoking motor. Having to push the saw can also be the sign of a dull blade, and a dull blade is a dangerous blade.

6. **Use both hands.**

REMEMBER

Your right hand is working the saw, but your left hand arguably plays a bigger role in making an accurate and safe cut. It should be doing two things at the same time: pressing down on the workpiece to hold it steady and prevent it from moving and, simultaneously, helping keep your saw on a straight line as it cuts. Many pros use their thumb and forefinger on the back edge of the saw's *shoe*, the flat metal plate that glides over the work surface to keep the saw moving in a perfectly straight line. The straightedge of a square also makes an excellent saw guide (as seen in Figure 5-4), keeping your fingers well away from the saw's blade (see more on squares later in this chapter).

TIP

A good clamp can hold the board securely to your sawhorse. Just make sure it's not placed in the direct path of the saw.

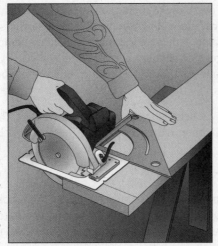

FIGURE 5-4:
A speed square makes a handy straightedge for your circular saw to follow.

7. **Finish strong.**

   As you reach the end of the cut, the weight of the wood will make the scrap end want to pull away, sometimes breaking off before the cut is complete. To avoid this, have a helper gently support the scrap end to keep it from free-falling. If you're flying solo, speed up during the final half-inch or so of the cut with a quick little push. Let go of the trigger and allow the blade to come to a full stop before pulling the saw away from the work-piece.

Mastering the miter saw

A miter saw (also known as a chop saw) is even easier to use than a circular saw. You won't need to adjust the blade's depth, because it's designed to be pulled down to a factory-designated stopping point. Stabilizing the board is still critical; the saw's metal fence and bed (which you can see in Chapter 3) help hold the work-piece steady and allow you to concentrate more on making the cut. Slight pressure on the good end of the board is still necessary during the cut itself. Start the blade early and keep it just on the scrap side of your cut line.

REMEMBER

You shouldn't try to "chop" when you're using a "chop saw." Despite the tool's catchy nickname, your motion should be considerably slower and more deliberate than the word *chop* suggests. Squeeze the trigger and allow the blade to reach full speed before contacting the wood. Then, use a slow and steady pull to bring the blade all the way through the lumber.

Miter saws are set up to cut straight cuts, the kind you'll need most often. But they can also be adjusted to cut any angle from 1 to 89 degrees. Most saws have factory-set stops every 5 degrees, but can also be manually manipulated and set at any angle needed. On most models, it involves simply squeezing a release lever and rotating the blade assembly until a pointer lines up with the desired angle on a numbered miter guide, then locking the blade assembly in place. Check your particular saw's user's manual for instructions on how to change the angle of the saw blade.

WARNING

Because of the way it's made, the blade of a miter saw spins *toward* you. This means that when the teeth of the blade touch the wood, you may feel the saw pull toward you as well. Keep your fingers out of the path of the blade and a good grip on the handle to maintain a stable saw and make a safe cut.

TIP

When cutting dense pressure-treated lumber (or hardwoods like oak) with a miter saw, many contractors get halfway through the cut and then pull the blade up away from the wood momentarily, with the blade still spinning. This allows the blade to get back up to full speed and ensures that the second half of the cut goes as cleanly and smoothly as the first half.

Ripping lumber

Crosscutting, or cutting across, a piece of lumber is relatively easy, whether you use a circular saw or a miter saw. But sometimes you may need to make a long cut that runs parallel with the long edge of the lumber to get a narrower or thinner piece of wood. An example would be cutting a 10-foot-long 2x4 in such a way that you end up with a 10-foot-long board that's still 1½ inches thick, but only 2⅞ inches wide. This is called *ripping* a board, and it's a bit trickier, because you're cutting *with* the grain of the wood and usually for a much longer distance than when crosscutting.

Keeping the cut straight is usually the hardest part of a rip cut. Unless you're a very skilled carpenter, free-handing it is generally not a good idea and will result in a wavy, ragged cut — and that's if you manage to complete the cut without binding the blade trying to keep it in a straight line!

Mark both ends of the board at the appropriate spot, and connect the marks in one long straight line with either a pencil and straightedge or a chalk line. Next, clamp a metal straightedge or another long piece of lumber in a position that will serve as a guide for your saw's plate while allowing the blade to cut along the line. Proceed with the cut using Steps 1 through 7 listed in the previous section, "Using a circular saw."

Pay special attention to how you stabilize the work-piece. The nature of a rip cut means that there's often precious little lumber to support the plate of your circular saw as it does its job. If needed, use another piece of lumber the same thickness as the one you're ripping to give your saw more surface area to glide across during the long cut.

TIP

Despite your best efforts, slowly and carefully guiding a powerful circular saw all the way down a rip cut that can easily be 6, 8, or 10 feet long may prove to be too much for you. If this is the case, try performing a rip cut with a jigsaw. It's much lighter, less bulky, and the up-and-down nature of the blade's action can make it easier to maneuver through a long and precise cut. It will likely take longer than using a circular saw, but one perfect cut that takes 5 minutes is better than wasting three pieces of lumber trying to get it done in 30 seconds.

REMEMBER

If you own or have access to a table saw, this is one of the things it's best at. Just be sure to follow all safety precautions, because table saws can be exceptionally dangerous, especially in the hands of a first-time user. Circular saws, jigsaws, and table saws all have the ability to perform an angled rip cut as well, where the blade doesn't slice through the wood perfectly plumb. Consult your saw's user's manual on how to adjust the blade for an angled cut, and consider practicing on a piece of scrap wood before attempting an angled rip cut on an important piece of coop lumber.

Cutting sheet goods

Using a handheld circular saw to cut large 4-x-8-foot sheets of plywood can prove challenging, even for master carpenters. If you're ripping a full sheet, you're almost guaranteed to have to stop the saw at least once mid-cut to reposition yourself and/or your grip on the saw. Having a locked-down straightedge to use as a saw guide is a lifesaver.

TIP

Consider measuring the distance from your saw's blade to its shoe plate and then clamping a long piece of scrap plywood that far off your cut line, as shown in Figure 5-5. Now the left edge of your circular saw's shoe plate can glide up against the scrap for the length of the cut. Just make sure you use the factory edge of the scrap to ensure a truly straight line.

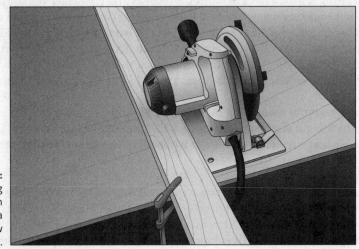

FIGURE 5-5:
Use a long piece of thin scrap as a straight saw guide.

When you use a circular saw to cut sheet goods, you should follow the basic steps that we provide in the earlier section "Using a circular saw safely." The following tips should help reduce splintering, a common problem when cutting plywood. For subfloor decking that will be covered, rough and splintered edges may not matter much. But for exterior sheathing or T1-11 paneling that will be visible in the finished coop, these hints can really improve the overall look with crisp, clean, professional-looking lines (see Chapter 4 for more information on these sheet goods).

>> **Put your best face down:** When cutting sheet material with a circular saw, the carpenter's adage is "best face down." This means measuring and marking the back of the sheet of plywood, and positioning the saw so that the best side — the side that will be exposed upon installation — faces down during cutting. Because of the way the blade spins, its teeth will crash upward through the topmost plane of the plywood. That results in splintering on the same side that the saw is running on. By having your "best face down," you're keeping that side as clean as possible.

REMEMBER

The "best face down" rule applies only to circular saws. A table saw's blade is oriented differently; its teeth will come down through the piece instead of up. Therefore, with a table saw, the opposite is true: "best face up."

>> **Score!:** Plywood tends to splinter easily, even if you're using a proper plywood blade (see the earlier section "Selecting saw blades" for more about this type of blade). After you establish your cut line, try running over it with a sharp utility knife (use a straightedge!). This breaks the topmost layer of fibers and reduces splintering during the real cut.

TIP

Some pros like to use their circular saw to score a piece of plywood. To do this, unplug your saw, set the blade at a depth so that it makes just a shallow cut in the top surface of the wood, and then plug it back in and make your first pass down the cut line. Next, readjust the blade to cut all the way through on the second pass. Making two full passes with the saw is more work, but it's a great way to get nice, clean plywood cuts and perfect your sawing skills in the process.

>> **Tape it off:** A strip of masking tape over the cut line keeps the wood fibers from tearing and breaking during the cut. Be sure to really affix the tape to the plywood so it doesn't peel away and gum up your saw blade. And remove the tape immediately after the cut, when it's more likely to release cleanly and splinter-free.

Assembling Materials

Pounding nails and sinking screws may seem as self-explanatory as it gets. But you'll likely be spending more minutes on these tasks than any other phase of your coop build, so any insider info that makes them go more quickly and easily, with better results and fewer bumps and bruises, is worth spending a moment on, don't you agree?

Nailing it down

For many people, the hammer was the first tool they ever used, taking a few labored swings as a youngster under the watchful eye of dear old Dad or a kindly grandfather. Unfortunately for some, their skills haven't improved much. There's actually an art to properly and efficiently wielding man's most basic tool. Ideally, it should take no more than five or six swings to fully drive any nail. These hammering hints will help you nail down solid technique:

>> **Get a grip:** Good hammering starts with how you grip the tool. Many beginners tend to grab the hammer close to the head or in the middle of the handle. But to get the most smash from each swing, you should grip the hammer at the end of the handle, as shown in Figure 5-6. Now you're using leverage to do a good portion of the pounding for you.

>> **It's *not* all in the wrist:** In an attempt to minimize sore muscles later, many a rookie will use short, tapping strokes that work the wrist while keeping the arm itself pretty still. This is a great way to wear out your wrist, but not an efficient way to drive a nail. Use the whole forearm, swinging from the elbow and keeping the wrist locked straight (see Figure 5-6). This puts even more of the hammer's weight to use.

>> **Thave your thumbs:** Thufferin' thuccotash! Saturday morning cartoon characters would have you believe that smashing your thumb until it turns bright red and swells up to the size of a basketball is just the inevitable result of using a hammer. Not so. The fingertips of your nonhammering hand do need to be in harm's way as you get a nail started, but after you've gently tapped it into an upright position, their job is done. Delete those digits from the danger zone, using additional force and velocity with each swing and making sure the hammer's face is parallel with the nail's head as it strikes.

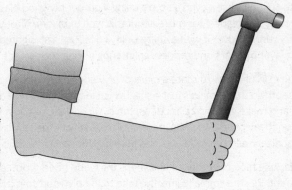

FIGURE 5-6: Grip a hammer at the end of the handle and use the whole forearm to swing.

TIP

Using a tiny nail that's too small to grip without beating your fingertips bloody? Try holding it tightly with needle-nose pliers or some thin cardboard until that nano-nail is standing on its own. A cheap plastic comb also does a serviceable job of holding a nail between its teeth while you get it started.

>> **Pull with power:** Pulling nails can be like, well, pulling teeth, especially with long nails that max out the claw end's reach. Sliding a scrap block of 2x4 under the hammer's head (as seen in Figure 5-7) can give you the leverage you need for prying that nail free. It also protects the finished surface of the wood around the nail.

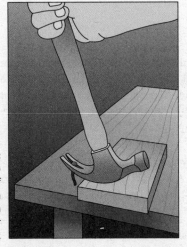

FIGURE 5-7: Use a scrap block of wood to gain leverage on a nail slated for removal.

Screwing it in place

Forget forearm muscles, hand-eye coordination, and elbow grease. Correctly driving a screw is all about the tool. Most cordless drill drivers let you select the amount of *torque* — or twisting power — you use to do the work. Dial up a higher torque setting (as shown in Figure 5-8) when you need more driving power, like when you're doing some heavy-duty framing or screwing into a really thick piece of lumber like a 4x4. If your screws are going too deep or you're stripping out your screw heads, select a lower torque setting. There's no magic formula for selecting a torque setting; just adjust as you work until you find a setting that adequately drives the screw without stripping the head or burying the screw too deep.

TIP

If you've been using one torque setting with success and suddenly find yourself needing more power to sink a screw, try a fresh battery in the drill.

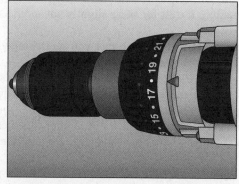

FIGURE 5-8:
Dial up more or less torque to sufficiently sink your screws.

PUTTING A STOP TO SPLIST ENDS

Splitting a piece of wood with a nail happens to even the best carpenters. You can't save the piece of wood when it does happen, so you have to start over with measuring, marking, and cutting a new piece. But there are some tricks you can use to keep it from happening again.

- **Blunt the tips:** Turn your nail over and tap the pointed end until it's blunt. A sharp nail tip acts like a wedge, splitting the fibers of the wood apart as it's driven. A dull tip crushes the fibers, making a split less likely.

- **Downsize:** Use the smallest nail you can that still gets the job done. A thinner nail is less likely than a thick one to split a piece of lumber.

- **Lube:** Coating a nail with beeswax or petroleum jelly, or even rubbing it against a bar of soap helps lubricate the nail, helping it to glide through the wood fibers as it's driven and minimizing the friction that can cause splitting. (This is a great trick to use on screws, and makes them easier to remove, too!)

- **Pre-drill:** Use a drill bit that's smaller in diameter than the nail and bore a pilot hole in the wood before nailing through the hole.

- **Try new wood:** Wood dries out over time, becoming more prone to splitting when nailed. If you're using some old lumber that was lying around in your basement and find it splitting, try a new piece from the lumberyard. Fresh lumber has more moisture in it and lets nails glide through more easily.

Joining Pieces at Tricky Angles

Even if you use all of carpentry's top techniques and trickiest tricks with your nails and screws, you'll still find occasions during your coop build where you have to think outside the box. This often occurs where two pieces of lumber meet at an unusual angle, or when you don't have an easily-accessible surface through which to drive a fastener by normal means. Maybe it's a tight corner where your hammer simply won't reach. Perhaps it's up on the roof, where your 2x6 rafter rests against the 2x4 cap plate. Or say you need to add a stud to a wall that's already framed and in place. Whatever the situation, it calls for some slightly advanced skills.

When the best way to join two particular pieces together leaves you scratching your head, try wrapping your head around these techniques instead.

TIP

If you're having trouble figuring out how to make a particular connection, it may be worth paying a visit to your local home improvement center. Near the lumber aisle, you should find a wide variety of galvanized brackets designed to hold pieces of lumber together in all kinds of odd positions. (Commonly used joist hangers and hurricane clips are described in detail in Chapter 7.)

Toe-nailing

Toe-nailing is the practice of driving a nail through the end of one board into another board at an angle. This is a necessary skill to master if you ever want to add a vertical stud into a section of wall that's already standing. It can be difficult to get the hang of right away, and often looks messy and sloppy when it's done, but with a little practice, the ability to toe-nail is a handy weapon to have in your coop-building arsenal.

TIP

You can perform this technique with screws, too, whereby it's usually called — as X-rated as it may sound — *toe-screwing*. The steps are the same.

Follow these steps (and refer to Figure 5-9) to properly toe-nail:

1. **Choose a longer nail than you've been using.**

   Those 2½-inch-long 8d nails may work just fine for assembling that section of wall framing (as described in Chapter 7), when you can drive them through the top and bottom plates and into the ends of the studs. But if you add a stud to a wall that's already standing, you have to hammer through the ends of the stud into the plates at a diagonal of between 45 and 60 degrees. That means you need a longer nail. Bump it up to a 16d nail (3½ inches long), and you should be good to go.

TIP

   When in doubt, hold a nail up to the outside of the joint to visualize how it will penetrate the plate once you sink it (as shown in Figure 5-9a). You want at least 1 inch of the nail to end up in the plate itself.

2. **Brace the backside of the stud.**

   It's imperative that you provide some strong reinforcement to the backside of the stud you'll be pounding against. Many pros use their foot to hold the stud steady (thereby driving the nail just a few inches in front of their actual toenail).

An even more foolproof method is to use a scrap piece of wood as a spacer block (as shown in Figure 5-9b). Cut the scrap to the exact distance between the wall studs and hammer to your heart's content, knowing that your spacer block is keeping the stud in perfect alignment.

Once you've finished toe-nailing, the spacer block may be wedged pretty tightly between the studs. Try fastening a simple handle to the top of the spacer block, which should allow you to wiggle it free to use again.

3. **Start the nail straight.**

If you try to drive the nail at a 45- to 60- degree angle from the beginning, it will almost undoubtedly slide down the vertical stud as you hammer. To get around this, start the nail into the stud perfectly horizontal (as shown in Figure 5-9c). Tap it in about a quarter of an inch.

4. **Tilt the nail and drive it in at an angle.**

Use your fingers to pull the nail up to an angle just past 45 degrees (slightly more vertical than horizontal), as shown in Figure 5-9d. Still holding the nail, give the nail a few light taps to get it started in this new direction. Then take your nonhammering hand away and pound normally, driving the nail at its new angle until the head hits the side of the stud.

5. **Don't go too far.**

Because the flat nailhead will meet the side of the stud at an angle, a portion of the head will likely be sticking out a bit from the stud. Continuing to beat on the nail to fully sink the head into the wood will only serve to bend the nail and possibly split the end of the stud.

6. **Toe-nail in 3s.**

Typically, if you're toe-nailing a stud in place, you'll want three nails per stud end. Go with two nails through one side, then a third on the backside (in the center, between the other two nails) to help kick the stud back into position (see Figure 5-9e).

Many fledgling DIYers find it helpful to drill pilot holes in the stud before toe-nailing. Bored at the proper 45-to-60-degree angle, they help guide the nails through the stud, even if your hammering skills are a little suspect.

Pocket hole joinery

For locations where toe-screwing is your best option but you're not wild about unsightly exposed screw heads, consider *pocket hole joinery*. This technique, a favorite among experienced carpenters and furniture-makers, takes a toe-screw and shrouds it in a larger angled hole (see Figure 5-10). The "pocket hole," technically called a *counterbore*, is wide enough to contain the entire screw — head and all — and leaves a clean surface with no hardware visible. Best of all, it makes a super-tight, rock-solid joint.

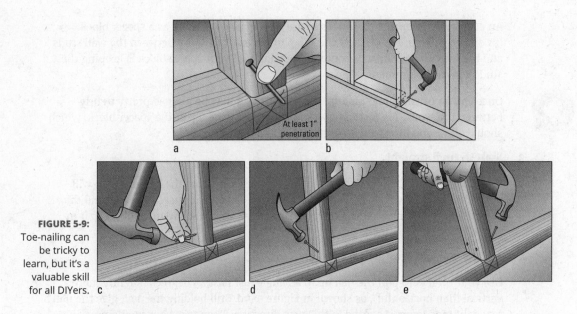

At least 1"
penetration

a   b

FIGURE 5-9:
Toe-nailing can be tricky to learn, but it's a valuable skill for all DIYers.

c   d   e

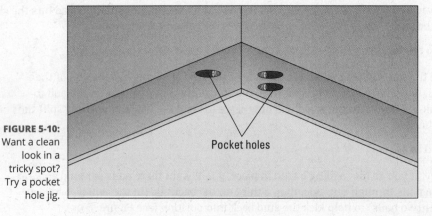

Pocket holes

FIGURE 5-10:
Want a clean look in a tricky spot? Try a pocket hole jig.

Only the most skilled of woodworkers would attempt this kind of connection freehand. For the rest of us tool hounds, there's a gadget that makes it as easy as operating a drill. Jigs manufactured by the Kreg Tool Company (recognized as the leader in the pocket hole niche) feature metal guides that allow the DIYer to drill a perfect counterbore at just the right angle and depth while simultaneously drilling a pilot hole for the screw. These kits have reasonable starting prices and provide ultra-professional results.

Plates

If all else fails in fastening two pieces of wood to each other, pay a visit to your local home improvement store and wander the hardware aisles. You'll probably find some sort of metal connector that you can adapt to your needs. Mending plates, angle plates, and tee plates (as seen in Figure 5-11) can be lifesavers in all sorts of building applications. It's probably a good idea to have a selection of them handy, along with screws that fit the plates' holes and are appropriate for the thickness of the lumber you'll be working with.

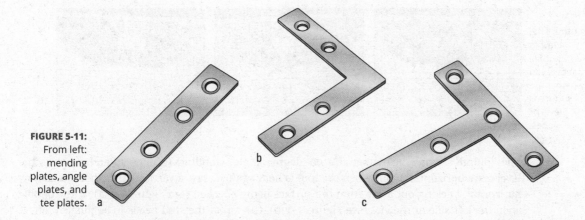

FIGURE 5-11:
From left:
mending
plates, angle
plates, and
tee plates.

a

b

c

Using Levels and Squares

Levels and squares are among the most straightforward tools you'll use in your coop build. (Want to check for level? Grab your level. Need to confirm that a corner is square? Reach for your square. If only all tools were so obviously and simply named.) But a surprisingly large number of homeowners don't really understand how to read a level or interpret a square. Even more don't realize how truly versatile these tools are and how many other uses they have.

The following sections describe what you need to know to get the most out of these everyday tools.

Carpenter's level

Whether you're using a small, pocket-sized torpedo level, a long, 4-foot carpenter's model, or something in between, they all work on the same principle. And just as with the tape measure (described earlier in this chapter), knowing how to "read" what the tool is telling you is a critical skill for any DIYer to learn.

Interpreting the bubble inside the vial of a carpenter's level is easy. If the bubble is perfectly centered between the lines marked on the vial (as shown in Figure 5-12a), you're dealing with a level surface. Pat yourself on the back and move on.

But what if the bubble isn't centered between the lines? Yes, it means "not level," but a tilted bubble also tells you at a glance where your problem is. The bubble, because it's air, will always rise to the top. So on a horizontal board, for example, if the bubble tilts to the left of the hash marks (as shown in Figure 5-12b), the left side of the board is the high side. You need to either drop the left end of the board or raise the right end. If the bubble "reads" right (as shown in Figure 5-12c), the opposite is true: Raise the left side or drop the right side.

Most levels have more than one bubble vial, so you can check a horizontal surface for "level" or a vertical surface for "plumb." The "level" bubble vial is almost always in the center of the tool; it's the one that runs parallel with the tool itself and sits horizontally when placed on a horizontal surface.

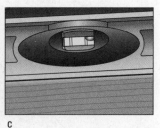

FIGURE 5-12:
A level's bubble shows you whether the surface is level or slopes.

a   b   c

The "plumb" bubble vial is set at a 90-degree angle, standing straight up and down as the level rests horizontally. But when the tool is held against, say, a vertical stud, that vial is now horizontal, sticking out away from the surface being checked (see Figure 5-13a). If the bubble is on the left side of the vial (see Figure 5-13b), the top of the stud needs to be pushed left. If the bubble floats to the right (see Figure 5-13c), the top of the stud needs to go that way, too.

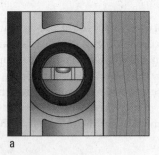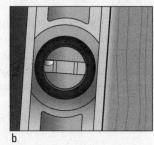

FIGURE 5-13:
Most levels also show plumb, or which way a vertical member is leaning.

a   b   c

TIP

Any level you purchase at your local hardware store will come precalibrated from the factory. But it's easy to check the accuracy of a level (even a brand-new one) before you start building. It's also a good idea; just a few hard drops on the ground can knock one or more of the level's bubble vials out of whack, resulting in skewed readings. To check a level, simply rest it on a flat surface and note the precise placement of the bubble. Now spin the level 180 degrees so the right end of the level becomes the left end and vice-versa; check the bubble again. It should be in the exact same spot. If it's not, get a new level. (This works even if the surface you're using isn't actually level.)

A level also makes a nifty straightedge, which makes it a versatile tool that you'll reach for again and again. It can be held or clamped in place on a piece of wood and used as a saw guide for cutting straight lines. Or use it for just marking long lines: Measure and mark two spots, then connect the dots with the level, and run a pencil alongside it for a precisely straight line that's more reliable than a length of wood. Some levels even have convenient tape measure markings printed on one edge to minimize the time you spend going back and forth from tool to tool.

Specialty levels

As mentioned in Chapter 3, you may find that having more than one type of level is handy during your coop build. Line levels and post levels are extremely common, especially for setting posts. (That phase of a larger coop build is tackled step by step in Chapter 6.) Both types of these specialty levels, however, work the same way as the carpenter's level described in the preceding section.

>> **Line level:** A line level is basically a single bubble vial that hangs from a string. It can check the tops of a line of fence posts, for example, to ensure that they're all sitting at the same height. Read it the way you would a carpenter's level: A left-leaning bubble means the left side is high, and a bubble rising to the right means the right side is high; a centered bubble is the ultimate goal.

>> **Post level:** A post level is an L-shaped bracket that checks two adjacent sides of a post for plumb at the same time. A long rubber band can be wrapped around the post for hands-free operation. Some post levels feature magnetic strips so they can be used on metal posts, too. A post level reads just like a regular level, only now, you need to read both bubbles simultaneously and adjust as needed until both show plumb.

Speed squares

A speed square (which, oddly enough, is actually triangular in shape) establishes a perfect 90-degree angle and a precise 45-degree angle. This makes it a valuable tool for marking straight lines (like when marking a cut line across a 2x4, as described earlier in this chapter) or scribing a true 45-degree angle from a single starting point.

But a square can do so many more things, too, making it one of the most helpful tools you'll carry in your tool belt. (See Figure 5-14 for a basic anatomy lesson.) For example:

>> Squares have incremental markings along their edges like a tape measure. If you need a quick, basic measurement (most squares don't get more detailed than quarter-inches), it's handy in a pinch.

>> As mentioned earlier in this chapter, the speed square makes a fantastic cutting guide for a circular saw. The thick body of the square gives the saw's shoe plenty of material to ride against as it works through the cut. Most pros like to hold the lip of the square against the board edge farthest from them, in essence, pulling the square toward them with one hand as they push the saw through the board with the other hand. This helps hold everything very steady and reduces the chances of something (the saw, the square, the board, your hands) slipping mid-cut.

>> Most speed squares have a series of notches in the body (usually near the ruler markings) that can help you scribe a long, straight line parallel to the edge of the board. Need to rip 1 inch off of that full-length 2x4? Find the notch that sits 1 inch off the edge and position your pencil in it. Now slide the square's lip along the entire board, marking the board as you go. Voilá.

>> A square also makes checking right angles a breeze. Want to make sure two boards meet at precisely 90 degrees? Slap a square inside the corner. If the corner of the square doesn't touch the inside corner of the joint or if the square doesn't meet both edges simultaneously, it's not 90 degrees. Or hold it against the outside of a corner to make sure the edges line up with the square.

>> A speed square also features protractor-like markings along its long, 45-degree side (that's the hypotenuse, for you closet geometry buffs). This allows you to read or mark any angle from 1 degree to 89 degrees. Four out of our five coops in Part 3 require cutting some pieces of lumber at 15-, 30-, or 60-degree angles. There are even some 21- and 34½-degree cuts! The markings on the long side of a speed square help you lay out any of these angles with ease and precision.

Here's how to lay out and mark a sample cut of 30 degrees:

1. **Mark one side of the wood.**

   Instead of marking a straight line across the piece of lumber, measure and mark just one edge, where you want your angle to begin.

2. **Set the pivot point.**

   Place the lip of the speed square against the wood. Slide it so that the tiny notch at the 90-degree corner of the square — called the *pivot point* — meets your marked edge.

3. **Swing the square.**

   Holding the square's pivot point tightly against the wood edge at the marked spot, swing the square until the desired angle number (in this case, 30) lines up with the edge of the board.

4. **Mark the angled line.**

   Still keeping the square firm against the board, mark a line down the square and across the surface of the wood, starting at the pivot point. Cutting along this line will give you a 30-degree angle.

TIP

Every speed square made by the Swanson Tool Company (inventors of the tool) famously comes with a detailed booklet explaining all of these uses and more, including how to use the dizzying array of numbers and lines stamped into the tool to tackle more complicated functions, like laying out and cutting roofing rafters.

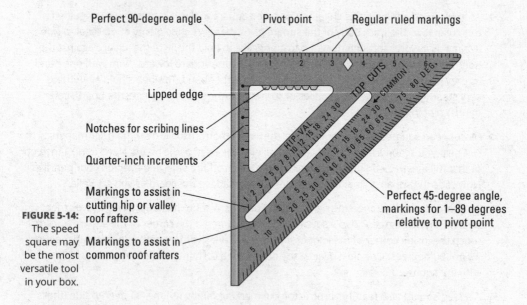

Perfect 90-degree angle

Pivot point

Regular ruled markings

Lipped edge

Notches for scribing lines

Quarter-inch increments

Markings to assist in cutting hip or valley roof rafters

Markings to assist in common roof rafters

Perfect 45-degree angle, markings for 1–89 degrees relative to pivot point

FIGURE 5-14: The speed square may be the most versatile tool in your box.

2

Constructing
a Coop

Chapter **6**

Preparing the Site

I n a perfect world, every chicken-owner would have a sprawling expanse of flat, level, unused ground just ready and waiting for a newly-constructed coop. Most of us, however, live in the real world, not a perfect one. And that means that the spot you envision building your coop on may currently be occupied by something else — a neglected flower bed, perhaps, or the kids' kickball diamond. Or maybe the spot you've chosen *would* be perfect . . . if only it weren't on a steep slope. In any case, you probably have some prep work to do before the real building begins, so roll up your sleeves and dig in.

Prepping the site for your chicken coop isn't glamorous work (or even much fun, truthfully), but it's the first important step toward a successful build. Building a chicken coop is like anything else in life: If you don't start with a solid foundation, whatever you build on top of it just won't last.

This chapter tackles getting your build site ready for a coop — clearing it off, ensuring that it's level, and sinking timber posts in concrete or employing another means of supporting the frame. If you're not sure where to place your coop, flip to Chapter 2 for guidance on selecting a site.

First Things First: Clearing the Site

The best spot for a chicken coop is usually a bare patch of earth. But of course, any patch of earth can *become* bare with time and elbow grease, along with the garden tools we describe in Chapter 3. Follow these guidelines:

>> **Clearing out ground cover:** If all you need to do is clear out some leaf litter or mulch, a heavy-tined rake should fit the bill. If you want to remove some lawn, a square-edged spade or flat-bladed shovel can cut the sod into strips and slice underneath the roots.

>> **Getting rid of existing plants:** Any plants, flowers, bushes, or shrubs that are within the footprint of the coop will have to go. Use a shovel and/or a mattock to dig these out, roots and all.

>> **Removing miscellaneous barriers:** If you have trouble keeping your footing while walking the site, so will your chickens. Dig up monster tree roots and heavy rocks with a mattock.

Checking the Level of the Ground

If the ground isn't level and you start building anyway, don't be surprised if you step back from your finished coop to find something that looks like it's right out of a Dr. Seuss book, with crazy angles that lean this way and that. You can check the *grade* (levelness of the ground) now and avoid a nasty surprise later.

To use a line level, follow these steps:

1. **Tap a few stakes into the ground around the perimeter of the site.**

   Leave most of each stake exposed. These stakes are temporary helpers only, and ramming them firmly into the earth will only make them harder to pull out. Tap them in just enough so that they stand upright and can take some light side-to-side tugging. Plus, the more of the stake you leave exposed, the more room you have to work with as you determine how level your grade is.

2. **Tie the string to one stake; extend the string to a second stake.**

3. **Pull the string taut and slide the line level onto the string.**

4. **Raise or lower the string against the second stake until the bubble reads level (see Figure 6-1).**

5. **Repeat the process for the other stakes.**

   You can use the same string, tied around the first stake, one at a time on each of the other stakes for quick reference. If you'd like to take a step back and survey the entire site at a glance, use a new string at every stake, keeping them all tied at the same level (or as close as you can get) on the first stake.

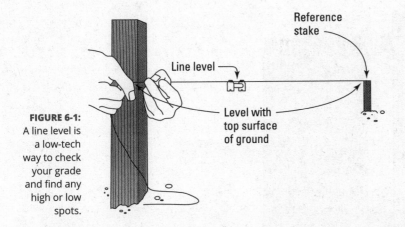

FIGURE 6-1:
A line level is a low-tech way to check your grade and find any high or low spots.

TIP

A line level will help you determine the overall slope of your build site, but the string also makes it easy to spot any particularly high or low spots within the footprint of the site. If you've got a severe bowl effect where water could pool, you'll see it using the string as a reference. Now's the time to shave down the high spots with a shovel and use the soil to fill in any low spots.

Installing Posts in the Ground

If your coop location is severely sloped or you simply want to elevate the coop off the ground, you'll need to build the coop on vertical support posts. Think about how a deck on the back of a house is most often constructed: on big, beefy timbers anchored in the earth. Instead of making the ground level, you create level posts to build upon.

Lumber isn't meant to withstand long-term, wood-to-soil contact. In some landscape applications, even pressure-treated timbers that are buried directly in soil will start to rot within seven years. Therefore, driving a support post into bare earth is not a recommended practice.

A better alternative is to encase the end of that wooden post in an underground column of concrete on top of a bed of gravel. This encasement is called a *footing*, and it will last from 10 to 15 years. Figure 6-2 shows a common way for DIYers to think about footings.

A proper concrete footing provides solid, weight-bearing support for your coop. But before you start mixing up concrete, you need a hole to pour it into . . . you dig? You also need braces to keep your posts upright in the concrete. We explain this process (called *sinking posts*) in the following sections.

Digging holes for footings

REMEMBER

For 4x4 posts, dig holes about 8 inches in diameter. This allows for plenty of concrete on all sides of the post and makes for a sturdy support base.

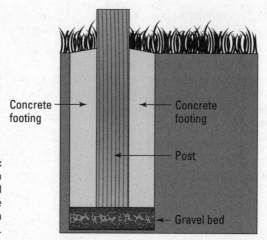

FIGURE 6-2:
Cross-section of a typical concrete footing with a 4x4 post.

Concrete footing → ← Concrete footing

Post →

Gravel bed ←

How deep should your footings be? That's actually a deep question. The short answer is: It depends. It depends on your climate, your soil type, the size and weight of the coop you're building, and more. In some areas, burying a shallow pad of concrete is sufficient. In other locales, the hole for that footing has to extend all the way to truly solid ground or below the *frost line* (the depth at which groundwater freezes) to prevent *heaving* (buckling of the ground to the point that the whole footing pops out of the ground during a cold snap). In some places, that means digging down over 5 feet! Your safest bet is to check with your local building inspection office to find out what the codes in your area require. At the very least, consult with a trusted contractor — someone in the deck-building, construction, or landscaping field who digs footings regularly — to see what is generally accepted in your area.

The following sections describe two methods for digging holes: using a posthole digger and using a power auger (Chapter 3 introduces both tools).

REMEMBER Few things will ruin your day faster than plunging a steel shovel blade into a buried power line or underground pipe. Call local authorities a few days before you begin digging; they'll come to your property and clearly mark all utility lines for you. You can find the number in any phone book, or call 811 from anywhere in the United States to be connected to a local office.

Using a posthole digger

For most weekend warriors, digging a hole in which to sink a post means spending some time with the good old posthole digger. It's relatively easy to get the hang of, most of us have one hanging in the garage or could borrow one from a friend or neighbor, and it costs only your time, sweat, and energy to use. The downside? Digging postholes can be exhausting work. If you have more than a few to dig or if your ground is particularly hard, this step may wipe out the better part of a build day, and wipe you out in the process.

Whether your posthole digger is an ancient, wood-handled clamshell digger that's been buried in the toolshed for years or a spiffy new model with fiberglass handles, the concept is the same. Ram the pointed blades straight down into the ground and pull the handles apart to squeeze the blades together, pinching a pile of soil between them. Lift the tool out and bring the handles back together to empty the dirt out. Repeat. Over and over. And over.

Be aware that the deeper you dig, the wider the top of the hole may get. That's because the handles may start hitting the walls of the hole when you spread them apart. A post-hole digger designed with scissor-hinged, offset handles combats this common problem. It works the same way, but allows you to dig deeper without also unintentionally digging wider. And the offset handles keep you from banging your knuckles together as you work.

Using a power auger

If you have many footings to dig, a power auger may be worth renting. You use one by starting the gasoline-powered engine with a pull string and holding on tight while the giant steel bit spins itself into the ground. If you've never used one, get the full lesson from the rental shop on how to operate the tool safely, and then take a few minutes once you get home to practice with it in a far corner of your yard. But be forewarned: Just because this is a power tool doesn't mean it's effortless to use. It's extremely heavy and very powerful. In fact, after a day running a power auger, you could find yourself with more bumps, bruises, and aches than if you had used a posthole digger.

If you're just talking about a hole at each of the four corners of your coop site, renting a power auger almost certainly isn't worth the extra expense and hassle. It makes sense only if you have a lot of deep holes to dig, say, for a very large, free-standing chicken run. Then, it could be a lifesaver. Here are several things to keep in mind if you decide to go the power auger route:

>> **Teamwork:** For 8-inch-wide holes to accommodate 4x4s, rent a two-man auger. As the name implies, you'll need a buddy's help. It has two sets of handlebars and weighs more than you think. It'll weigh even more once the spiral bit gets loaded up with soil during the dig.

>> **Torque:** A power auger produces a wicked amount of torque. Be prepared for the kickback that revving the throttle will unleash. A power auger can easily knock you right off your feet, so brace your legs against the handlebars and hold on tight.

While digging, it doesn't take much of a root or rock to jam the spinning blade and bring it to a sudden, jolting stop. You may end up using that posthole digger after all just to chop through whatever's down there.

>> **Depth:** Reaching your desired depth may require an extension rod. To use one, you'll have to physically dismantle the auger bit from the head and then reattach it with the steel extension in between, as shown in Figure 6-3. The extension rod will immediately make the auger much heavier to lift and more unwieldy to operate.

As you pull a spinning auger out of its hole, some loose dirt will undoubtedly fall right back in. To make sure you really dig to the depth you want, auger down a few inches past your target depth, so when loose dirt spills back in, you're still in good shape. Or use a posthole digger instead to scoop out the loose soil that falls into an augered hole. Either way, remember to measure the finished depth of your hole before moving on to be sure you're not too deep or too shallow.

Don't try to start a new hole with the extension rod already in place, as the auger will simply be too tall to control. (It could be over your head!) Drill down to the maximum depth of the bit itself; then pull out and add the extension to finish off the hole.

FIGURE 6-3:
An extension rod allows for deeper holes — and makes for a heavier auger.

REMEMBER

>> **Pace:** Slow and steady is the name of the game. Plan on lifting that spinning auger out of the hole frequently to let the dirt slough off. If you try to dig the entire hole in one marathon push, it could become too overloaded with soil to lift out. Even worse, it could corkscrew itself right into the earth and get stuck. Getting it out under these circumstances is brutally difficult and time-consuming, making your power auger not quite the timesaver you had hoped for.

Bracing your posts

With the holes dug, it's time to drop in your posts. Don't worry about the length now. Keep them all longer than you need (finished post height according to your plans + depth of post hole + some extra); you'll trim the tops later. A few inches of gravel in the bottom of each hole helps keep ground moisture from puddling around the base of the post and should be the first thing to go in the hole.

Now set the post on the gravel. Because the hole is so much wider than the post, the post is unlikely to stand upright on its own. You'll need braces to temporarily hold it for you. Normally made from scrap wood or inexpensive lengths of 1x2 or 1x4 stock, a typical brace consists of two pieces:

>> A small stake driven into the ground to anchor the whole thing.

>> A second piece of wood that's long enough to be fastened to both the post and the stake. Attach it to the stake with a screw or nail, but don't fasten it all the way down. Leave it loose so that it swings, as if hinged.

You'll use two braces for each post, as shown in Figure 6-4.

Then use a post level (see Chapter 3 for more about this tool) and adjust the post until it's standing perfectly plumb. Once it is, secure the brace to the post. It's a good idea to start a screw into the end of the brace so that you'll be able to drive it into the post itself quickly and with a minimum of effort, helping ensure that the post remains plumb while you work.

If you're using a nail gun, the task is even easier, because everything is instantly held fast with one quick squeeze of the trigger. Using a regular hammer and nails isn't recommended for bracing, because the act of hammering a nail into the post usually knocks the post right out of plumb. (And after the concrete hardens, that nail will be harder to remove than a screw.)

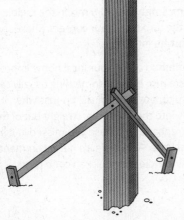

FIGURE 6-4:
Braces made from scrap wood will hold your posts plumb while you prepare your concrete.

TIP

Keeping a heavy post steady and plumb while simultaneously using a hammer and/or a screw gun to fasten braces to it is easy — if you're an octopus. Get a helper; having an extra set of hands and eyes to assist you saves time and eliminates a lot of frustration.

Mixing and pouring concrete

Your posts are in the holes, braced in the plumb position. Time to encase them in concrete. Just as there is more than one way to skin a cat (although it's never been adequately explained why one would *want* to), there are also varying methods for mixing concrete.

>> **Using simple hand tools:** Empty the contents of a bag of concrete mix into a wheelbarrow. Add water according to the package directions. Mix with a hoe. It doesn't get much simpler than that. It takes a surprisingly small amount of water to get the right consistency, so mix thoroughly before adding water. If you do add more water, do so sparingly — a few quick squirts at a time. You're aiming for gloppy, not soupy.

>> **Employing cool gadgets:** There are a few accessories out there designed to make mixing concrete a little less taxing. Some people swear by their favorite; others find them all more trouble than they're worth. It's up to you to gauge how much elbow grease you're willing to exert on mixing concrete, and whether items like these would help you significantly:

- **Drill mixer:** Think about the electric mixer in your kitchen. A drill mixer (see Figure 6-5) works the same way, only it's bigger. This long, drill-powered, paddle-shaped bit can be used to mix wet concrete in a 5-gallon bucket or a deep wheelbarrow. Just like your kitchen mixer, though, it can make a royal mess, sending your ingredients flying everywhere. And thickening concrete can provide a lot of resistance, so a typical cordless drill may not have enough oomph to do the job.

- **Odjob:** A handy tool with a funny name, the Odjob (see Figure 6-6) is basically a big bucket with a screw-on lid. Dump in a bag of concrete mix, add in one lid's worth of water (cleverly designed to hold just the right amount), close it up tight, and roll it around on the ground for a minute. The bucket has baffles inside that mix the concrete and water together as it rolls.

- **Cretesheet:** Working on the same principle as the Odjob, the Cretesheet (see Figure 6-7) is a heavy tarp with handles on the corners. Empty your dry mix in the middle, add water, and then get a friend to help you pick up the tarp by all four corners. Shake the tarp to mix the ingredients; use it to pour the concrete into place.

TIP

» **Not mixing at all:** This ranks as the easiest method of all. Your local home improvement center sells, right alongside its regular concrete mix, bags of something usually called "quick-set" or "fast-setting" concrete. It is the ultimate in no-muss, no-fuss when it comes to concrete. Simply dump the dry mix directly into the post hole, straight out of the bag, and then add water on top. It mixes itself. I'll repeat that in case you missed it: *It mixes itself!* Whoever invented this stuff deserves a Nobel Prize. And while it's not recommended for super-heavy-duty applications, it can certainly handle your chicken coop posts.

FIGURE 6-5:
A drill mixer can whip up a batch of concrete if you don't mind the mess.

FIGURE 6-6:
The Odjob uses internal baffles to mix the concrete as you roll it.

WARNING

Mixing concrete (using any method) kicks up a lot of dust, which is not something you really want to breathe in. Wearing an inexpensive dust mask while you're opening bags of it, dumping it out, and mixing it up is smart.

FIGURE 6-7:
The Cretesheet is a tool-free way to mix concrete and pour it where you want it.

After you mix concrete with any of the preceding methods, pour it into your holes with a shovel. But be careful! The top of your post is braced, but the bottom isn't. If you throw shovelfuls of wet, heavy concrete mix into one side of the hole, it can cause the bottom of the post to move and shift out of plumb. Add concrete a little at a time around all four sides of the post to keep the weight distributed evenly. You may even want to keep your post level attached as you fill the hole to make sure everything stays where you want it.

REMEMBER

Concrete needs time to harden, or *cure.* While the wet mix should hold the posts in place almost immediately, don't be in a rush to remove those braces. And by all means, don't jump right into hammering and sawing on those posts until the concrete has had a chance to set. You'll just end up knocking the posts out of plumb and causing major headaches for yourself down the road. Give wet concrete a bare minimum of 24 hours before you start messing with the posts. Forty-eight hours is even better.

Securing Posts Aboveground

Surrounding posts with concrete in holes isn't the only way to establish a base for the frame of your chicken coop. Alternatively, posts can be mounted aboveground on top of concrete footings through the use of metal post bases, or concrete pier blocks can serve as bases for your posts.

Mounting posts on top of concrete footings

Although a post encased in a concrete footing will last longer than one placed directly in the ground, eventually it will draw moisture out of the concrete itself and rot. To get the maximum life out of your posts, you can pour a solid footing, and then mount your post on top with a metal post bracket (see Figure 6-8). This bracket remains aboveground and accommodates a 4x4 post, which is adequate for most coops.

The instructions for digging holes and mixing and pouring concrete are the same as those given in the earlier section "Installing Posts in the Ground." However, the process for bracing the posts differs somewhat.

The metal post base is secured to the concrete footing with a heavy bolt, which is inserted either into the concrete while wet, or into a drilled-out hole in the concrete after it has hardened. (Using metal post bases also translates to lower lumber costs, because you're not paying for all those extra feet of wood to be buried below ground.)

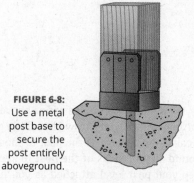

FIGURE 6-8:
Use a metal post base to secure the post entirely aboveground.

Using concrete pier blocks

Versatile, pyramid-shaped concrete pier blocks (see Figure 6-9) are recessed at the top to accept a 4x4 standing on end; some can also accommodate 2x4 joists. When placed on a level concrete footing pad or solid ground, they can offer a sturdy support base for the frame of your coop without the need to dig holes, brace posts, and use concrete.

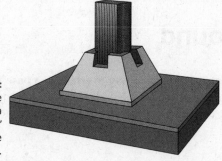

FIGURE 6-9:
A concrete pier block is a great "no-dig" way to elevate your coop.

Be aware that these precast pier blocks rely on the weight of the structure alone to hold them in place, because there's usually no convenient way to permanently fasten lumber to them. Furthermore, the blocks must be used on level ground or the entire structure will be unstable.

TIP

Some precast pier blocks have a metal post base anchored into the top. If your home improvement center doesn't stock these, they may be able to order them for you. Or it may be worth a trip across town to pick some up at another store if you want to use pier blocks but prefer the added security of mounting your posts to them.

Topping Your Posts

However you set your support posts, they're likely sticking up out of the ground at different heights. Before going any further, you need to level the posts, or *top* them. Follow these steps to top your posts:

1. **Establish the finished floor level of the coop.**

   Mark this spot on one post as a reference point.

REMEMBER

   When determining the level of your finished floor, don't forget to take into account the thickness of the flooring material itself! For common, ¾-inch plywood, come down this same amount on each of the posts as you mark and level them.

2. **Transfer that same level to each of the other posts.**

   With a helper and a line level, mark the other posts at the finished floor height. (See the earlier section "Checking the Level of the Ground" for details on using a line level.) Scribe a line around all four sides of each post with a pencil and a square.

3. **Cut your posts.**

   If you've encased your posts in concrete, you'll have to make your cuts with the posts standing in position. This is done most easily with a circular saw held sideways (see Figure 6-10). Depending on how deep your saw cuts, you may have to go around every side of each post — and even cut through the center of the post with a hand saw.

   If you're using pre-cast pier blocks or metal post bases and can still remove the posts, simply take each 4x4 to the saw and make the cut. Replace the trimmed posts, and you now have a level base on which to build your coop.

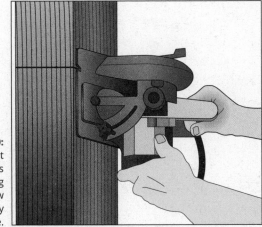

FIGURE 6-10:
Topping a set post requires carefully using a circular saw at a tricky angle.

IN THIS CHAPTER

» **Constructing a solid subfloor**

» **Framing a basic stud wall**

» **Allowing for doors and windows**

» **Raising and fastening walls**

» **Raising the roof**

Chapter **7**

Laying the Lumber: Framing 101

R emember building with Lincoln Logs or Legos as a kid? Taking those individual pieces and assembling them together in just the right way to construct a rustic cabin for your Star Wars Stormtroopers or a multicolored barn for your toy ponies? That's framing: piecing together structural members to create a shell that, in this case, will be the frame of your chicken coop.

With some coops, this frame is just a skeleton and will be completely covered with some kind of solid exterior sheathing, like plywood or siding. (More on this step is found in Chapter 8.) Other coops, like the all-in-one that we introduce in Chapter 2, wrap a portion of the frame in welded wire and use it as the run. This kind of coop leaves your framing abilities on display for the whole world to see.

But framing isn't necessarily a secret skill that requires hiring a high-priced carpenter. It may not be child's play, but even the most novice DIYer can construct a well-built frame using some basic techniques, employing a few trusted tricks of the trade, and going rogue with one or two clever shortcuts. This chapter covers creating a floor structure to build on, framing exterior walls, adding doors and windows, and putting a roof over your coop.

Building a Subfloor

If your coop is meant to house chickens and chickens only — meaning you physically can't or won't enter it at any time — you don't need a heavy-duty floor. In this kind of coop (like a small A-frame or tractor) the "floor" of the chickens' quarters is basically just a platform. The floor still needs to be supported in some basic way that keeps it level and prevents it from sagging or buckling, but it doesn't have to be built with a carefully-spaced series of bulky 2x6 joists underneath. Think of it as shelving; it doesn't need to support anything heavier than the chickens.

But if you're looking at a walk-in design, a *subfloor* is the first thing you'll need to build. As the name suggests, a subfloor is what's below the flooring surface that you'll actually walk on. It's the structural makeup of the floor. In the following sections, we explain the two basic steps of building a subfloor: framing joists and adding decking.

Framing the joists

Your subfloor will, in essence, be a large box that's the exact footprint of the coop itself. The box is typically made up of two opposing *rim joists*, two of the beams that will compose the subfloor, and a series of joists. The joists are shorter boards that complete the perimeter of the frame and then also fill in the box at regular 16-inch intervals. (See Figure 7-1 for a detailed look at the pieces that go into a subfloor.) The following sections provide tips for assembling the subfloor frame.

FIGURE 7-1: A typical subfloor layout.

Attaching rim joists

Begin by fastening two rim joists to your posts, the tops of which should all be level with one another. (Advice on leveling post tops is found in Chapter 6.) You have two options:

>> The easiest way to attach rim joists is to use a metal post-to-beam connector, which is nailed or screwed to the top of a post and accepts a beam on top (see Figure 7-2a). These connectors have holes already spaced at all the proper locations; drive your appropriately-sized

fasteners of choice through each of the holes into both the post and the beam. (See Chapter 4 for help determining the length that your nails or screws should be.)

>> Alternatively, notch your posts to accept a beam, and connect them with nuts and bolts that go through both pieces of lumber (see Figure 7-2b).

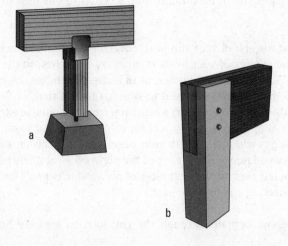

FIGURE 7-2:
A metal post-to-beam bracket (a) and a notched post-to-beam con-nection (b).

TIP

Notching a post requires some detailed measuring and tricky cuts with perhaps more than one saw from your arsenal. If you make a mistake, you may ruin that piece of lumber — and if that post is encased in a concrete footing, you've got a problem. Unless you're supremely confident in your carpentry skills, keep it simple: Buy the metal brackets.

WARNING

Don't be tempted to just use a few nails or decking screws to fasten a rim joist directly to the outside of a 4x4 post. Sure, it's cheaper than buying metal connectors and quicker than notching a post, but it's a recipe for disaster. Here's why: The perimeter of the subfloor will end up having to support the most weight, because it bears the load of the walls and the roof. Using a metal bracket or notching a post to accept a beam properly transfers a portion of that weight to the post, ensuring that your subfloor stays put.

Adding joists

Finish the frame's perimeter with two joists. These should stretch from the end of one rim joist to the matching end of the other rim joist and complete the square shape. Make sure that the two pieces of lumber are level on the top and square at the corners. Then drive three nails or screws through the rim joist into the end of the joist, spacing the fasteners evenly down the height of the joist.

Now is also a good time to confirm the squareness of your frame. Hook a tape measure over one corner of the frame and stretch it diagonally across the frame to the opposite corner. Make a note of that measurement. Then measure the other diagonal. (If you were to do this with two tape measures at the same time, they would form an X.) If your frame is truly square, the two measurements will be identical.

TIP

If your frame isn't quite square, you may be able to whack it into square with a heavy hammer. Just knock the frame one way or the other until the measurements are the same. (If you're more than a couple inches off, stop everything. Double-check all your boards' measurements and figure out what went wrong; this can't be fixed later.)

Add more joists inside the frame, keeping them parallel with one another and spacing them so that they're 16 inches apart *on center,* meaning from the center of one joist to the center of the next.

The first floor joist will sit at the end of your rim joist, covering the first 1½ inches. From the second joist on, you want the centers of your joists to sit every 16 inches, so the nails in your rim joist will be driven into the centers of the joists, at 16 inches, 32 inches, 48 inches, and so on. Because your joist is 1½ inches wide, you need to subtract half of that, or ¾ inch, to find the measurements to the ends of your joists. Start measuring from the outside edge of your end floor joist. Make your marks at 15¼ inches, 31¼ inches, 47¼ inches, and so on. As you place and fasten the joists, their edges will line up with your pencil marks. (This ensures that when you lay that first sheet of plywood flooring, the edge of the plywood — at 48 inches — will rest on the fourth joist with room left over for the next piece of plywood to begin. This allows ample room to nail through your sheeting and into the joist.)

Use three evenly-spaced screws or nails through the rim joist to securely hold each joist in place.

TIP

Although many perfectly good coops are built on joists that are 24 or even 36 inches apart, 16-inch spacing is standard in most carpentry projects: floors, walls, roofs, and so on. Because many building materials are sold in sizes that already assume 16-inch spacing, it's probably easiest to stick with that measurement. You'll never have any issues if you space your joists 16 inches apart; a floor built on joists spaced farther apart could eventually sag due to inadequate support.

If you like, you can use metal brackets instead. Called *joist hangers* (see Figure 7-3), they hold the joists to the inside of the rim joists and keep all of the nail and screw heads inside the frame, where they're hidden from view. Some rookie DIYers like using joist hangers if they're concerned about driving fasteners through the rim joist and actually hitting the center marks of the joists. Hangers eliminate that guesswork. Most also feature a pointed prong that you can hammer into the rim joist to hold the floor joist in alignment for you while you work. That's a plus for many DIYers who aren't comfortable holding the joist in perfect position with one hand while they drive fasteners with the other. The downside to joist hangers, however, is that buying joist hangers (two per floor joist) ends up costing quite a bit more than simply using nails or screws.

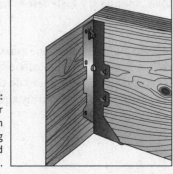

FIGURE 7-3:
A joist hanger keeps the rim joists looking clean and professional.

REMEMBER

Working from the outside to secure a joist, the fasteners must go all the way through the rim joist and well into the joist itself. A 3-inch nail or screw is most commonly used. But if you use joist hangers, your nails and screws are much shorter, 1¼ inches at most, so they don't penetrate all the way through and leave pointy ends sticking out of the exposed frame.

Installing the decking

With the subfloor frame constructed, it's time to add something you can walk on: the decking. For 99 percent of all chicken coops, this is made from sheets of ¾-inch plywood. (Refer to Chapter 4 for more on choosing the right type of plywood.)

Pieces of plywood or OSB (either full sheets or cut-to-size pieces, depending on how big your coop is) are put down on top of the floor joists and secured with nails or screws. Be sure that as you're sinking your nails or screws, you're driving into the floor joists below the plywood; some people find it helpful to snap a chalk line across the plywood at each joist location, creating a handy nailing guide that tells you exactly where to place your fasteners (we explain how to use a chalk line in Chapter 5).

TIP

A store-bought sheet of plywood or OSB can also help you check your frame for square. Line up a factory-cut corner, a perfect 90 degrees, on a corner of the frame. Does the frame line up exactly with the edges of the plywood? If so, you're golden. If not, get out that hammer and tap the frame one way or the other until all edges are perfectly aligned — and, therefore, square.

TECHNICAL STUFF

Many carpenters run a bead of construction adhesive (like Liquid Nails) along the top of each joist before nailing or screwing the plywood down. This provides added holding power, minimizing squeaks in the finished floor. For chicken coop purposes, this probably isn't necessary. (Will you really be spending enough time inside your chicken coop to be bothered by a stray creak here or there?) But if you have a caulk gun handy and want the added insurance, go for it with a construction adhesive rated for outdoor use.

Your first nails or screws through a piece of plywood should be in two outer corners to hold the piece securely in place. (One in each corner is sufficient for now.) You may want to step back and double-check everything before moving on. Removing two nails to make an adjustment is easier than waiting until the entire sheet is nailed in place.

Pop in a new nail or screw every six inches along all the edges and every joist. Keep the pointed tips of the fasteners ½ inch or so away from the wood's edge (to prevent splitting) and angle them slightly so they're being driven toward the center of the joist below.

REMEMBER

Make sure the heads of your fasteners are completely flush with the plywood, or even dimpled slightly within the wood's surface. Nobody (including your chickens) wants to get snagged on a nailhead poking up out of the floor.

If you need to abut two pieces of plywood together, both edges must land on a joist so that the ends of the plywood are supported from below. (Carpentry nerds say that this is where the pieces *break*.) You have to measure and cut your plywood carefully, because you have just 1½ inches — the width of the joist — to play with. (Making that joint break at the exact center of the joist is optimal.)

It's a good idea to leave a slight gap between the two sheets of plywood. Moisture and temperature changes will force the wood to expand. If the two pieces are slammed together tightly, the swelling wood has nowhere to go but up, causing a hump in the floor. Leaving a gap as small as just ¹⁄₃₂ inch (as shown in Figure 7-4) will allow for that expansion to occur, but won't be noticeable to you or your chickens while walking on the floor.

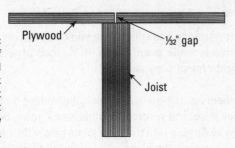

FIGURE 7-4:
Sheets of plywood should break on a floor joist with a slight gap to allow for expansion.

Plywood

¹⁄₃₂" gap

Joist

Want a no-brainer way to measure out a ¹⁄₃₂-inch gap? Forget squinting at those tiny markings on your tape measure all day long. Use an upright credit card as a spacer. Just remember to put it back in your wallet before your next trip to the hardware store for supplies.

You might be tempted at this stage to really "floor it" and add a final flooring material like linoleum or wire mesh (both popular options that are discussed in Chapter 4) now, before moving on with coop construction. There's really nothing stopping you from doing so, but understand that when you erect the coop's walls, you'll be installing them over the top of this extra flooring layer. That may make the walls harder to hold steady as you fasten them down. You'll also have to drive your fasteners through this additional material. And if you ever want to change or replace the flooring material, you'll have to cut it out, because several inches of it will be sandwiched underneath the coop walls. All things considered, it's probably easier to wait until the coop itself is built before adding a top layer of flooring.

Framing the Walls

Depending on which coop style you decide to build, much of the advice given thus far in this chapter and in Chapter 6 may not apply to you. Perhaps you don't need to do any real prep work on the site. Maybe you're not anchoring your coop into the ground on concrete footings. It could be that you don't need a big walk-in coop, and therefore don't need a floor.

But every chicken coop has to have walls of some kind. For a large walk-in, the walls will be structurally similar to the ones in your home. For an all-in-one coop-and-run, they may be scaled down significantly. And if you're constructing a small A-frame or tractor coop, your "walls" may turn out to be little more than the sides of a large wooden box.

Walls provide the shelter for your chickens: shelter from the elements and from predators. It's important, then, to build them to be sturdy and to last. In the following sections, we describe the layout of a wall's studs and explain how to assemble a wall on the ground.

Laying out studs

At its most basic, every wall, regardless of size, is made up of a series of vertical members commonly called *studs.* The length, number, and exact layout of the studs can vary, but they provide the necessary framework or skeleton for any walled structure.

Studs (most often 2x4s but sometimes other sizes, like 2x3s) run straight up and down and must be held in place between horizontal pieces at the top and bottom. These pieces are called the top and bottom *plates.* The bottom plate gets secured to the floor (whether it's plywood, like the one we describe earlier in this chapter, or another material), while the top plate supports the roof overhead.

Doors and windows alter the layout of some studs (we review them in detail later in this chapter), but studs are almost always spaced 16 inches apart on center. The anatomy of a basic stud wall is shown in Figure 7-5.

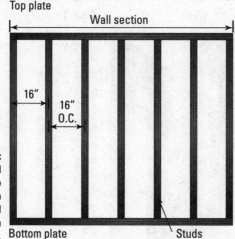

FIGURE 7-5: A typical stud wall, with top and bottom plates, as well as 16-inch stud spacing.

REMEMBER We know we just started talking about wall studs, but it's not too early to also start thinking about your coop's roof! Here's why: Many coop designs call for a *shed roof,* where the roof is one large plane that sits atop the coop and tilts toward one side, usually the back. (You can find lots more on roofs later in this chapter.) If that's what you're building, it generally means that at least one coop wall must be made quite a bit taller than the others. That's good to know now, while you're measuring, marking, and laying out the studs. If you build all four walls to the same height, that roof won't have much slope to it!

TIP As you're figuring out the placement of your studs, adding by multiples of 16 can be confusing for the mathematically-challenged. Check your tape measure: Many tapes feature some sort of special marking every 16 inches specifically for stud locations. This makes it easy to extend your tape and, at a glance, visualize studs at 0, 16, 32, 48, 64 inches, and so on.

WARNING If the wall you're laying out will be home to a door, a window, or some other opening (like an exterior-accessible nest box), these features will likely interfere with your stud spacing. Skip ahead to the section in this chapter titled "The extra parts needed to frame doors and windows" to get a sense of how to factor these openings into a stud wall before you proceed.

Assembling wall panels on the ground

If you've watched any home-makeover show on television, you've no doubt seen the moment when all the workers, in their matching hard hats and tool belts, raise an entire stud wall off the ground and walk it into an upright position, usually to an off-camera cheer from a crowd of bystanders.

Besides being the ultimate money shot of renovation TV, assembling a wall panel on the ground and lifting it into place is just a good building practice. Working on the ground lets you lay out the entire wall on a solid and level surface, easily access and adjust every piece of lumber, and finally raise it to a standing position only when everything is truly ready.

But perhaps the best reason to construct your wall on the ground is the ease of nailing that doing so provides. With your studs and plates laid out at your feet, you can simply nail through the top of the top plate and the bottom of the bottom plate, placing two nails (or screws) into each end of each stud (see Figure 7-6). No 45-degree toe-nailing, which, as explained in Chapter 5, can drive even an experienced carpenter crazy and will never be as sturdy.

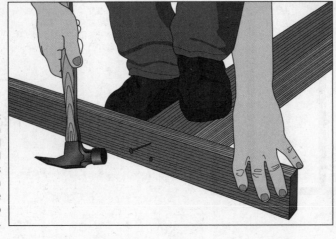

FIGURE 7-6: Assembling a section of wall on the ground allows nails to be driven through the plates into the studs.

Here are the steps for assembling a basic section of stud wall on the ground:

1. **Cut the top and bottom plates to the exact desired length of the wall.**

    REMEMBER

    For the first wall you build, you can make it the same length as the floor. For a wall that meets another wall, though, don't forget to factor in the width of the first wall. If you're using 2x4s, subtract 3½ inches (the width of one stud wall) from the length to ensure that you don't end up building a too-long wall that sticks out beyond the edge of the subfloor. If you're building a wall that will be placed in between two other walls, subtract 7 inches, the width of both stud walls. (These measurements change if you decide to use 2x3s or another board size.)

2. **Mark the plates at each stud location, maintaining 16-inch spacing.**

    To be sure that your stud marks on the top plate line up precisely with the stud marks on the bottom plate, most builders mark their plates in pairs. Turn the plates on their sides (with the 1½-inch side facing up) and sandwich them together, making sure the ends are flush with one another. Stretch your tape measure from one end of the plates and make your marks at 16, 32, 48 inches, and so on. Then, using a square aligned on each mark, scribe a line across both studs. (These will be the studs' centers.) Make sure that a stud closes off the far end of the wall frame, even if it falls short of the standard 16-inch spacing.

REMEMBER

Your first stud goes at the very end of the plate, where your tape measure was hooked during marking. That means the *center* of that first stud is actually ¾ inch from the end of the plate. And that will put your other studs' centers at 16¾, 32¾, 48¾ inches, and so on. Remember to line up the edge of each stud with its pencil mark to maintain proper 16-inch spacing.

TIP

With their studs' center locations marked, many DIYers like to also trace both sides of a scrap piece of stud lumber onto the plate. This shows the exact thickness of the stud — and provides a can't-miss "safe zone" on the outside of the plate to refer to when driving the nails or screws.

3. **Measure, mark, and cut the studs.**

The length of the studs is determined by the eventual height of the wall. But remember that the top and bottom plates help make up that wall height. Factor these pieces in by subtracting 3 inches from your overall stud length, because each plate is 1½ inches thick. For example, say that you want to end up with a wall that's 6 feet (72 inches) tall. The bottom plate will be 1½ inches tall. The top plate, too. So a stud that fits between these plates should measure exactly 69 inches.

4. **Loosely fit all the pieces together.**

Using the coop floor as a level work surface, orient the boards so they are turned on their sides (with the 1½-inch side facing up). Position the bottom plate closest to the outside of the wall where it will go, and the top plate toward the center of the floor. Lay out the studs in between; lining up the stud ends with the locations marked on the plates.

REMEMBER

This *dry-fit* (when the wall members are laid out in their final position but not fastened together) is your last chance to step back and make adjustments or changes. Take a moment to double-check all your measurements one final time, too. Any alterations from this point on will likely involve pulling nails and therefore take a lot longer.

5. **Use fasteners to assemble the pieces into one section.**

Most builders stand inside the framework of the wall, using one foot to step down on both the stud and the plate, holding the pieces with their body weight while they swing downward and hammer the nails home. If you're using screws, you may be more comfortable kneeling on the lumber while you work. Either way, use two evenly-spaced nails or screws at each end (top and bottom) of each stud.

WARNING

Be sure that you're sinking your nails or screws straight, so they go into the meat of the stud. If they're angled upward, you could end up driving one right into the bottom of your foot. This is especially important if you're using a pneumatic nailer, because you won't have time to figure out what's happening until it's too late.

Framing Doors and Windows

The directions in the previous section work great for basic walls. But if you erect four basic walls, you'll find yourself and your chickens with no way to actually get into the coop. You'll need some sort of door. And because lighting and ventilation are so important to your flock (as we discuss in Chapter 2), you'll probably want to add at least one window, too. In the following sections, we explain the extra parts you need for framing doors and windows and the extra steps you need to follow during the framing process.

The extra parts needed to frame doors and windows

While the actual door and window units themselves will be added later in the build (and dealt with in Chapter 8), you must take them into account now, during framing. That means, in essence, leaving holes for them within the network of studs and plates that make up a wall. But, of course, it's not as easy as just leaving big, gaping holes. In order to provide proper support and stability for doors and windows, extra framing members must be added within the studwork: Each door and window must be individually framed. Figure 7-7 shows a breakdown of the additional parts and pieces that must be added to a stud wall prior to installing a door or window:

>> **King stud:** The same full height as a regular stud, a *king stud* acts as a side support for a window or a door.

>> **Header:** A *header* is a horizontal piece that makes up the top of a doorway or window. Because some of the studs in the wall are missing (where the doorway is), the header provides structural support by carrying the load of the framing above it. It fits between two king studs and is supported from below by pieces called trimmers.

>> **Trimmer:** A *trimmer* (sometimes called a *jack stud*) is a vertical piece that supports one end of a header. You'll use two trimmers per door or window opening. Trimmers are shorter than full studs, and are always doubled up with a full-length king stud.

>> **Sill:** A *sill* (or *saddle*) is a horizontal piece used at the bottom of a window opening.

>> **Cripple:** A *cripple* is a short blocking piece that extends from a bottom plate to the underside of a sill, or from a top plate to the topside of a header. Think of cripples as ministuds; lay them out to maintain the same 16-inch on-center spacing as the rest of the studs.

Walking through extra framing steps

TIP

Even though it involves some forward thinking and advance planning, it's best to include door and window framing in the wall-framing phase of your coop project. Adding a door or window after framing can certainly be done, but you'll find it somewhat more difficult and probably much messier.

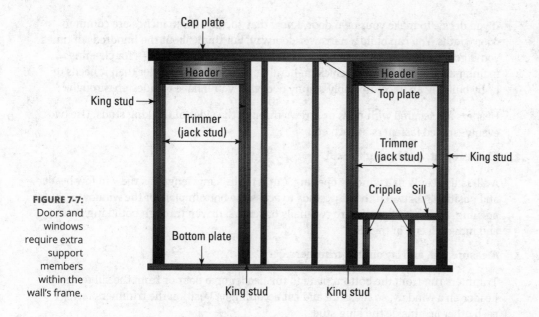

FIGURE 7-7:
Doors and windows require extra support members within the wall's frame.

Labels in figure: Cap plate, Header, Header, King stud, Top plate, Trimmer (jack stud), Trimmer (jack stud), King stud, Cripple, Sill, Bottom plate, King stud, King stud

To start framing the rough openings for a door or window, you need two key pieces of information: the door or window's measurements, and exactly where you want it to go. You'll then follow the same basic procedure as for a regular wall. Build on the ground, adding in a few extra steps as you go:

1. **Measure, cut, and lay out the plates, studs, and king studs.**

   The king studs will run the entire height of the wall (minus the top and bottom plates), so cut them to the same length as the rest of the studs on that wall. Position the king studs between the top and bottom plates, and fasten them with two evenly-spaced nails or screws through the top and bottom plates.

2. **Measure, cut, and lay out the header.**

   The header is the width of the final desired opening, plus the width of the two trimmers. For a doorway opening 30 inches wide, use a header that measures 33 inches. (After you add the trimmers at 1½ inches wide each, you'll be left with a 30-inch wide opening.)

   - If you're using a pre-hung door or window, use the outside measurements of the actual unit and add the width of two trimmers to arrive at your header length.

   - If you're making your own door or leaving a simple opening for a window (all covered in Chapter 8), use the measurements specified by your particular coop design.

   Position the header (a 2x4 or 4x4 is usually sufficient for a coop) in between the two king studs at the proper height on the wall, remembering to measure up from the outermost edge of the bottom plate (which will eventually be at floor level).

TIP

If you decide to make your own door, know that 30, 32, and 36 inches are common door widths. You can build a narrower doorway, but think about the hundreds of times you'll go in and out of the coop — often with a big armful of stuff — for cleaning, maintenance, or just general chicken-herding. Don't make it harder than it needs to be by building an uncomfortably skinny door that you'll have to sidestep through.

Headers are secured with nails or screws through the sides of the king studs. Use two evenly-spaced fasteners at each end.

3. **Measure, cut, and lay out the sill.**

   A sill is used only in a window opening. Cut it to the same length as the window header and position it between the king studs to act as the bottom plate of the window opening. Fasten it in place with two nails or screws, driven through both king studs and into each end of the sill.

4. **Measure, cut, and lay out the trimmers.**

   Trimmers run from the bottom plate to the header on a door or from the sill to the header on a window, so measure and cut accordingly. Position the trimmers facing each other just inside the king studs.

TIP

On a doorway, you may be tempted to cut out the section of bottom plate that runs between the trimmers now. After all, this is the doorway; that 2x4 along the floor will need to go, right? Actually, pros leave it in place until after the wall is up and secured. It helps keep the wall rigid as you lift and position it, holding both sides of the doorway in alignment. It's much easier to hand-saw that piece out later than to set a wall that's flexing as two independent sections.

5. **Measure, cut, and lay out the cripples.**

   Cripples are short ministuds that support a window sill and connect a door or window header to the top plate. When laying out cripples, keep them in line with the 16-inch spacing of the wall itself. But use an extra cripple to support each end of a header or sill.

6. **Use fasteners to assemble all pieces into a solid wall section.**

   Trimmers must be nailed vertically through the bottom plate like a king stud, but also nailed horizontally to the king stud itself to create a "double-wide" post. For cripples, nail through the bottom plate and sill or through the top plate and header as needed to secure cripples just like full-length studs.

Raising the Wall

Once the wall section is complete, getting it up off the ground and into place can be a real chore, particularly if it's a sizable wall. When big build crews do it, they all grab the top plate and lift on cue, stopping when the plate reaches their waist. Then, coordinating their movements to keep the wall from flexing too much and the weight distributed evenly, they raise it to their shoulders, and then walk toward the wall (as shown in Figure 7-8a), shuffling their hands down the studs as they go until the wall is standing upright (and being careful not to push it right over).

It's critical to get the placement of the wall exactly right before fastening it to the subfloor. Have your helper double-check its alignment with the subfloor frame, making sure the wall's outer edge sits flush with the subfloor's.

TIP

Snapping a chalk line or making pencil marks 3½ inches off of the floor's outer edge before you ever stand the wall up will give you a visual target to aim for with the wall. (Of course, that 3½ inches only works for a stud wall that's made of 2x4s. If you used 2x3s, your stud wall is a different width; your chalk line would need to be placed 2½ inches from the outer floor edge.)

Once the wall is upright and in position, have your helper hold the wall steady while you sink 3-inch nails or screws through the exposed bottom plate, two of them spaced evenly in each section of bottom plate that's exposed between a pair of studs (see Figure 7-8b).

FIGURE 7-8:
Teamwork makes raising and setting a wall easier.

WARNING

Don't nail through the bottom plate in any doorways! This short section of plate will have to be removed later. If it's not fastened to the subfloor, it can be cut out easily. If you secure it with nails or screws now, you'll just have to pry them up later.

For added holding power, many builders also add a toe-nail (or toe-screw) at every stud. (See Chapter 5 for tips on proper toe-nailing technique.) Be sure that each fastener goes through the stud at an angle that allows it to also penetrate the bottom plate, the subfloor decking, and the joist underneath.

WARNING

It's imperative that the nails or screws go through the subfloor and *into the joist below*. A wall that's secured only to a sheet of ¾-inch plywood or OSB may not stand up to a heavy gust of wind or a severe storm. Line up your fasteners about ¾ inch from the outside edge of the subfloor to make sure they dig into the joist under the plywood floor.

Fastening Walls Together

When two wall sections come together in a corner, they must be fastened to one another. Now, since you're really making visible progress at this stage, it's easy to get carried away and start nailing or screwing the two wall sections together so you can quickly move on. Hold on a minute, cowboy.

Your walls may be secured to the subfloor, but that doesn't mean they're set in their final positions. It's vitally important that the walls themselves be as square and as plumb as you can make them by following these steps:

1. **Check each wall section for square.**

   The easiest way to do this is to take two diagonal measurements across the whole wall. If the two measurements are equal, the wall is square. If they're off, use a heavy hammer to pound one end of the frame into alignment until the diagonal measurements are the same.

2. **Check each wall section for plumb.**

   Use a 4-foot level held vertically. You want the whole wall to stand plumb, but right now, just concentrate on the corner where it meets the other wall. If possible, have a partner check the other wall with a second level at the same time. The goal is for your two walls to meet each other perfectly — and remain plumb.

3. **Fasten the walls to one another.**

   When the walls meet cleanly and are plumb, drive screws or nails through the wider face of one stud into the narrower face of the other to hold the walls together (see Figure 7-9).

TIP

Have the screws or nails started and sunk almost through the first stud before aligning the walls together. This way, once you and your partner agree on plumb, one of you can quickly finish the attachment. If you're hammering nails the old-fashioned way, the other partner needs to hold on tightly to keep the walls plumb during all of that pounding, and be sure to check plumb again when you're through. Using a screw gun makes this step go much faster. Quickest of all, a pneumatic nailer means you need to hold the walls in position just long enough to tap the trigger.

FIGURE 7-9:
Fasten two walls to each other after you check that they're square and plumb.

Framing the Roof

Roofing can put a good scare into even an accomplished DIYer. A floor is no sweat. Walls are just giant rectangles. But a roof? That involves complicated geometry, crazy angles, and intimidating carpentry jargon like *trusses* and *rafters* and *pitched gables*, right? Well, yes and no.

The good news is that, because you're building a chicken coop and not a house, everything about the roof is on a much smaller scale. And in this case, "smaller" definitely translates to "easier." Framing a coop roof, even for a generously-sized walk-in, doesn't have to be any more complicated than setting a subfloor or whipping up a wall frame.

While a build-your-own coop can be as elaborate as you want it to be, it can also be as simple as your skills need it to be. Unless you're simply dying to dive head-first into advanced architecture and master-level carpentry, it's safe to rule out those complex roof styles you've heard contractors talk about and focus on the two simplest (as shown in Figure 7-10):

>> **Shed roof:** This is as basic as it gets. Picture a flat lid across the top of your coop. Now imagine lifting one end of the lid a little higher than the other. That's a shed roof, sometimes called a *pent roof* (see Figure 7-10a). Its single plane allows rainwater to drain off the roof, almost always taking runoff to the back of the coop.

>> **Gable roof:** A gabled roof is the classic, triangular roof shape (as shown in Figure 7-10b). It has two sides that start low and slope upward to meet in the middle at a long peak. This *ridge* runs the length of the building. It's like two shed roofs whose high ends are joined. To think of it another way, an A-frame is nothing more than a gable roof that rests on the ground.

In the following sections, we explain how to figure out the right pitch for your coop's roof, describe the anatomy of a roof, and walk you through the steps of building a roof frame with rafters.

FIGURE 7-10: The shed (pent) roof and the gable roof are common and easy to construct.

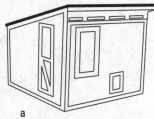 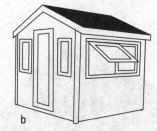

a   b

Getting the pitch perfect

A roof has to be sloped for rainwater to run off. In roof-speak, that slope is called the *pitch*. A steeply sloped roof that you wouldn't want to walk on is said to have a lot of pitch, while an almost-flat roof has very little pitch.

So what's the correct pitch for a roof? There is no magic number. In theory, it should be driven mainly by the weather. A roof in a dry area can get away with less pitch than a roof in a locale that receives heavy rain and snow. But in truth, a roof's pitch is determined largely by the desired look of the finished structure, the amount of materials needed to build it, personal preference, and even whether overhead clearances need to be factored in (low-hanging trees or power lines, for example). Typically, a steeper roof costs more to build and maintain, although for a chicken coop, the differences may prove to be negligible.

As a general rule, though, a coop roof should have a slope between 20 and 40 degrees. Less than 20, and you may experience problems with rainwater, snow, and excessive leaf litter collecting on the roof. More than 40, and the roof starts to get awfully steep and is a little extreme for a coop.

Maybe they do it just to show they're smarter than the rest of us, or maybe they just like having their own mysterious code. But roofers (at least in the United States) don't use degrees when they refer to a roof's pitch. They typically use a fractional notation that gives the number of inches of *rise* (or height) the roof gains per every 12 inches of *run* (or length). For example, a 4/12 roof (called "a four-twelve roof" or "a 4-pitch roof") gets 4 inches higher with every 12 inches of roof, and translates to a slope of precisely 18.43 degrees. An 8/12 ("eight-twelve" or "8-pitch") roof is pitched at a brainy 33.69 degrees. Imagine having to keep a measurement like *that* in your head for the entire project! (So maybe that explains why roofers don't talk in terms of degrees. . . .)

Analyzing a roof's anatomy

Just as a floor is supported by subfloor joists and the surface of a wall is backed by studs, the roof of any structure is built upon a skeleton of framing members (see Figure 7-11). The *rafters* are like the "joists" of the roof itself. They angle downward and are fastened to the top plates of your wall frame, generally on 16-inch or 24-inch centers. On a gable roof, the rafters meet at the top and are usually attached to a long ridge board. Horizontally-positioned ceiling joists and collar ties can be used to strengthen the rafters and create a true "A" shape, but they're only necessary in the largest of walk-in coops.

FIGURE 7-11:
The anatomy of a roof isn't all that different from a subfloor or stud wall.

Ridge board

Rafter

Cap plate

The classic A-shaped structures, when prefabricated, are called *trusses,* another fancy word thrown around by roofing professionals. Trusses go up quickly and are a huge timesaver in house construction. For a small building like a chicken coop, though, you're better off framing your roof from scratch, one stick of lumber at a time.

A home's roof rafters are typically 2x6 boards. After all, they need to be able to support a lot of weight — not just the plywood sheathing and shingles and such, but the occasional human being who goes up top to do maintenance, retrieve a Frisbee, or set up the annual "eight tiny reindeer" holiday display. You'll likely never need to get on top of your chicken coop, so 2x4 or 2x3 rafters will work just as well and save you a few bucks in the lumber aisle.

Building a roof frame

For a shed roof (with just one tilting plane), framing is pretty easy. During wall frame construction, you can simply build the front wall taller than the back wall, as shown in Figure 7-12. (You'll have to do the math to figure out how much taller to make it, factoring in the size of your particular coop and the pitch you want the roof to have.) Now you can use long studs to connect the front wall's top plate to the one on the back wall. How to make those connections is tackled in the later section "Cutting and attaching rafters."

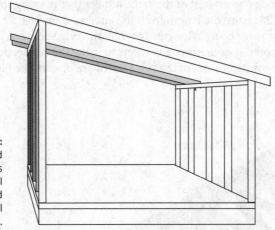

FIGURE 7-12:
Framing a shed roof begins during wall framing; build the front wall extra-tall.

Constructing a gable roof adds just a few steps:

1. **Run a second top plate around the perimeter.**

   Use new studs to create a double top plate, overlapping the corner joints of the existing top plates (see Figure 7-13) for extra rigidity and to help support the added weight of the roof rafters. To do this, cut a plate that is the same length as your first wall's top plate, plus the width of two more studs (to compensate for the two walls that attach to this first wall). You'll need two plates this length, atop opposite walls. Then measure and cut the plates for the other two walls.

   Fasten this second plate, often called a *cap plate,* with nails or screws. Because the wall's top plate already has fasteners at every stud, fasten the cap plate to the top plate in between studs (one fastener per gap) to avoid hitting the first set of nails or screws.

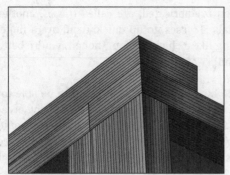

FIGURE 7-13: Overlap the cap plate ends for added strength.

2. **Install the center joists.**

   Position one stud plumb in the exact center of the top plate and fasten from underneath with two evenly-spaced framing fasteners, either nails or screws. Put a stud of the same height in the center of the opposite top plate. Make sure the tops are level with one another because these vertical members will support the ridge board that is the roof's peak.

3. **Install the ridge board.**

   Measure, mark, and cut a ridge board (out of the same lumber you're using for studs) to rest on the center joists. Measure the length from the outside of one center joist to the other to determine your ridge board's length. Toe-nail the ridge board to the center joists (as we explain in Chapter 5) or use metal brackets to make the connections. Use a small level to check that the vertical sides of the ridge board are plumb. See Figure 7-14 for what this should look like.

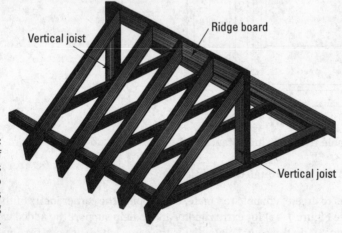

Vertical joist

Ridge board

Vertical joist

FIGURE 7-14: A gable roof frame starts with two vertical joists and a long ridge board.

Cutting and attaching rafters

A roof's pitch is what makes it so nerve-wracking for DIYers. Subfloor framing and wall framing are relatively simple tasks because you're fastening flat boards to other flat boards at nice, neat, perpendicular angles to one another. On the other hand, nailing one 2x4 to come off of another at a 20- or 30-degree angle can get even the most confident of weekend warriors quaking in his steel-toed work boots.

At the peak of the roof, the rafters must meet the ridge board at an angle. Then down at the cap plate, there's another angle to deal with. Professional roofers have all kinds of fancy tools to get this right: they bury their noses in little books that include headache-inducing grids and tables of numbers; they consult mathematical formulas etched into the sides of their speed squares; they even input their roof's measurements into specialized $100 calculators that then spit out the proper angles at which to cut their rafters.

The rafters do have to be securely fastened in order to safely hold the entire roof structure to the coop. But what if you just want to hang a handful of rafters one time for your little chicken coop without having to take a refresher course on the Pythagorean theorem? You can start taking eyeball guesses, but that's a surefire way to start wasting a lot of wood. Luckily, there's a totally low-tech trick for determining the rafter-to-ridge-board angle — and a way to avoid the tricky rafter-to-cap-plate "notch cut" altogether.

The brace-and-trace technique

TIP

Forget the mind-boggling arithmetic. Put away the protractor. Yes, the top of your rafter needs to be cut at an angle so that it sits flush against the ridge board. But to crack the code of what that angle actually is, you already have everything you need: a ladder, a pencil, and a key art skill that you learned when you were six. Here is the old-school, no-math, foolproof method known from this moment on as the "brace-and-trace":

1. **On top of a ladder, rest one end of a long rafter on top of the wall's cap plate.**

   Whatever lumber you're using for studs will work for rafters, too.

2. **Maneuver the other end of the board until the top corner is aligned with the top of the ridge board along its outside end.**

   This will leave a triangular corner of the rafter sticking out past the ridge board, as shown in Figure 7-15.

3. **Using one hand to "brace" the board in this position, use a pencil along the ridge board to "trace" a vertical line down the rafter.**

   This pencil line is your cut line. It's called a *plumb cut* because it's perfectly vertical. But most important, you now know it will line up perfectly with the ridge board.

4. **Before taking the board down, have a partner on the ground mark the other end of the rafter at the point you want it to stop after it has cleared the top plate.**

   This, called the rafter's *tail*, will be the overhanging end you will see sticking out above the finished wall (see Figure 7-15). Make the mark at whatever spot on the tail looks good to the eye; just a few inches will probably be sufficient.

Now you can cut the first rafter on the traced line and at the mark for the tail, and then use this rafter as a template to pre-cut every other rafter you need.

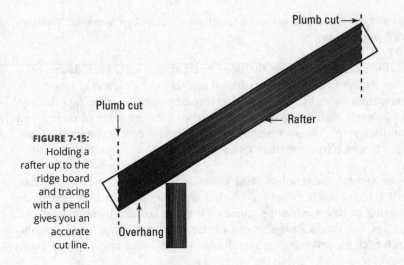

FIGURE 7-15:
Holding a rafter up to the ridge board and tracing with a pencil gives you an accurate cut line.

Rafter ties

Where the rafter tail meets the top plate of the wall, professional roofers and masochistic DIYers will attempt something called a "bird's-mouth cut." It's a triangular notch cut into the rafter so that it sits perfectly flush on the cap plate, hugging both the top and outer side of the cap plate. It looks nice and it's an impressive bit of carpentry, but you can achieve the same connection in a much simpler way.

A *rafter tie* (see Figure 7-16) is a metal bracket made for tying roof rafters and trusses to the framework of a wall. In some places, it's also referred to as a "hurricane clip," and as that nickname implies, its purpose is to reinforce the connections inside a home where super-high winds and tropical storms can rip a bird's-mouthed-and-toe-nailed roof right off its frame.

But it can be used by itself on the *outside* of a small utility structure (like a chicken coop) to attach a rafter to the cap and top plate without making any cuts at all to the rafter. Simply line up the rafter tie at the spot where the rafter and cap plate meet, and drive nails or screws into the tie's holes. Use two rafter ties per rafter for added strength.

FIGURE 7-16:
A no-cut rafter tie can secure a rafter to a wall.

Chapter **8**

Adding Walls, Doors, Windows, and a Roof

At this point in the building process, your coop is really starting to look like something. Getting it framed out (as described in Chapter 7) is a big step, but it doesn't leave you with a chicken-ready structure. For that, you need to encase that framework with solid walls, top it with a roof to keep the inside warm and dry, add working doors that allow easy access for you and/or your flock, and maybe even install windows that provide a way for sunlight and fresh air to get in and for stale, smelly air to escape.

These components come with their own unique considerations:

» Unlike the studs, the exterior walls will be on display in the finished coop, so style points start to count for a lot.

» Doors, opened and closed repeatedly, must stand up to a lot of use.

» Windows need sheets of glass or something else to cover the opening while still serving the necessary functions of a window.

» A roof has to keep out the wind and rain, because a leaky coop will prove to be problematic for both you and your birds.

But, as with every other phase of a coop-building project, there are plenty of easy-to-work-with materials to choose from, DIY-friendly methods to use, and time-tested tricks for getting it right. By the time you finish this chapter, that coop that was *starting* to look like something will be almost wrapped up and ready to welcome its flock home.

Putting Up Walls

When builders talk about "putting up a wall," they're usually referring to erecting the framework of studs and plates that we deal with in Chapter 7. But, while necessary, that open grid of lumber obviously won't do a very good job of keeping your chickens contained in their new shelter. You need to finish those walls off by cladding them in some sort of solid material.

Most DIYers pick plywood to do the job, especially on a small-scale, outdoor utility structure like a chicken coop. As detailed in Chapter 4, plywood is a readily found resource, available at any lumber store or building supply center. It's relatively inexpensive and quite strong and sturdy, yet easy to work with, even for beginner builders. But you'll find other options as well, and the following sections cover those, too.

Fastening plywood in place

Whether you're using regular ol' plywood or the dressier, grooved T1-11 paneling (see Chapter 4 for more on this), attaching your plywood to a stud frame is relatively simple. Follow these steps to ensure it's done correctly:

1. **Measure each wall to be covered.**

   Don't worry about subtracting anything yet for the door and window openings you framed in Chapter 7; it's easier to cut those out after the panel of plywood is up and in place. You're looking for the total height and width of each exterior wall.

REMEMBER

   If you have roof rafters to deal with, you may need to cut your plywood a little short of the full wall height. The downward angle of the rafter tail will often pinch a piece of plywood and prevent you from making it sit flush with the very top of the cap plate (that's the topmost framing member of each wall section, as described in Chapter 7). Subtract ⅛ inch, and the plywood should fit cleanly below the rafter tail and still be able to be fastened to the wall's top or cap plate.

2. **Consider how sheets will meet each other.**

   It's best if each exterior wall is just one solid sheet of plywood; with no seams to deal with. On a large coop, however, you'll have to do some planning where two sheets of plywood meet.

TIP

   Aim for vertically-running seams (two tall sheets standing side by side) as opposed to horizontally-oriented seams (where one piece rests on top of another). If moisture (like rain) ever gets into the seam, you want it to drain down the full length of the seam and run away. A horizontal seam can trap water, allowing it to either work its way into the coop interior or start rotting the wood that's trapping it. (If your coop walls are taller than 8 feet, there may be no way to avoid a few horizontal seams. Consider using trim to cover it on the finished coop.)

   When two pieces meet at a vertical seam, they must meet (or *break*) on a stud so that the edge of each sheet is fully supported by a backing surface that you can actually drive a nail or screw into.

REMEMBER

Anyplace where two sheets of plywood meet, leave a thin gap between the edges (as shown in Figure 8-1). This allows the wood to properly and safely expand due to moisture, temperature, and humidity. Most carpenters use a 16d nail as a temporary spacer to create a consistent ⅛-inch gap between pieces. (Don't worry; a gap this narrow shouldn't cause a real issue with rain or moisture finding its way into the coop. If the gap bothers you, use decorative trim to cover it; the plywood will still expand and contract as needed underneath.)

TECHNICAL STUFF

In Chapter 7, we suggest using a credit card to create a ⅟₃₂-inch gap between pieces of plywood or OSB on a floor. In this exterior wall discussion, we mention a ⅛-inch gap. Why the difference? As a rule, an exterior wall will be exposed to more moisture and wetness (rain, snow, humidity) than the interior of a floor. So wall pieces will probably swell more than those making up a floor. The slightly larger gap between wall pieces is simply to accommodate this additional swelling. Professional builders use 16d nails because they're plentiful and handy, never further away than the nearest tool pouch on-site. Leave a slightly narrower gap between the wall pieces of your chicken coop if you like; we won't tell. It shouldn't make a huge difference.

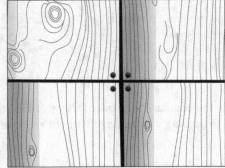

FIGURE 8-1: An expansion gap allows sheets of plywood to swell.

3. **Cut your plywood to the measurements of each wall.**

REMEMBER

Be sure to employ safe techniques for accurate cuts in sheet goods, as discussed in Chapter 5. And as the old saying goes: Measure twice, cut once. An "oops cut" in a full sheet of plywood often leaves you with a 4-x-8-foot piece of ruined scrap.

4. **Nail or screw the perimeter first.**

Working along the top and bottom edges of the sheet, drive one fastener into each stud. Working vertically along the studs that support the plywood's edges, use one fastener every 6 inches. Don't get any closer than ½ inch from the edge of the sheet to avoid splitting.

WARNING

Don't countersink the fasteners into the wood's surface! While driving the head of each fastener into the surface is the right way to ensure a level subfloor (as explained in Chapter 7), those little dimples will only mar a finished wall and leave unsightly pucker marks.

5. **Finish fastening the middle of the sheet.**

With the entire perimeter of a sheet secured, move inside (to what carpenters call the *field*) and nail or screw into each backing stud. Place your fasteners about one per 12 inches.

GOING AT IT ON THE GROUND

When building a home, many professional builders prefer to fasten their plywood sheathing to the framed wall before they raise the wall into a vertical standing position. If you're constructing a large walk-in coop, you may want to consider employing the same technique.

With the exterior side of the wall facing up, place your plywood sheets on top. Line up the corner of the plywood sheathing exactly on the corner of the stud frame using the factory edges of the plywood as a guide. Now make sure the stud frame under the plywood is square. If one side of the frame sticks out beyond the plywood, knock the whole thing into square with a hammer until it all lines up precisely. (This is a great trick for ensuring a wall's squareness without breaking out a tape measure!)

Most carpenters would rather work at their feet and drive nails or screws downward than work on ladders and hoist huge sheets of plywood up, trying to hold them flat against a vertical wall with one hand while hammering (sometimes overhead) with the other. Once the entire wall is sheathed at ground level, the wall section is raised using the methods described in Chapter 7.

Adding the plywood while the wall is still on the ground makes it significantly heavier to lift and put into place. So while working on the ground may turn sheathing into a one-person job, getting the wall off the ground will likely still require a helper or two.

Some types of exterior paneling, like the popular T1-11, have factory-milled tongue-and-groove edges that allow pieces to be snugly fit together without an expansion gap or visible seam.

Whenever possible, use the factory-cut edges where two pieces butt up to one another. Your expansion gap will be one consistent width, and you'll be left with a perfectly-straight line. If the visible seam bothers you, you can always cover it later with a piece of decorative trim.

Consider your coop's layout and floor plan before you get carried away with nailing up all the exterior walls. It would be a shame to secure that fourth and final wall, only to realize you now have no way to get *inside* the coop to start cutting out the door opening! Try a solid wall first to get the hang of putting up plywood. Then move to a wall with a doorway so you can still squeeze through the open studs of another wall to begin the process of cutting out the doorway opening (described in the next section). You'll now be able to use that doorway to get in and out of the coop as needed to do the final two walls.

Cutting out openings

While you can certainly factor in doors and windows right off the bat, most pros just sheathe over these openings with solid pieces of plywood to start with and then cut the holes out after everything is secured to the stud frame. Creating giant L-shaped pieces of plywood and fitting them together like a jigsaw puzzle around a doorway requires taking a lot of meticulous measurements and making several complicated cuts. And trying to pinpoint a window opening's exact location on a full 4-x-8-foot sheet, making the cut, and then having it all actually line up when you secure the plywood is a supremely tall order. If you miss by more than just a fraction of an inch, you may have just wasted an entire sheet of plywood.

To cut a door or window opening out of a piece of plywood that's already nailed in place on a stud wall, you'll probably have to make the cut from the *outside* of the coop, because the studs that frame the opening will prevent you from getting close enough with a saw to make a smooth, flush cut from the inside of the coop. But the whole process starts inside the coop:

1. **Drill a hole at each corner of the opening.**

   From inside the coop, place your drill bit as deep in each corner as possible. Drill all the way through the sheathing.

   If you'll be using a jigsaw or reciprocating saw to cut out the openings, use a drill bit that's bigger than the saw blade (see Step 3).

2. **Connect the dots.**

   Now, standing *outside* the coop, use a straightedge and pencil (or chalk line) to mark lines that connect the holes. This is the exact outline of your door or window opening.

3. **Make the cut.**

   Use a circular saw, jigsaw, or reciprocating saw to cut along the marked lines. Keep a steady hand and use a straightedge saw guide to avoid veering off and plunging the blade into the studs. (Flip to Chapter 5 for details on using saws safely and effectively.)

   If you're using a jigsaw or reciprocating saw, fully insert the blade into the hole to easily start the cut. With a circular saw, unlock the depth lever and start with the blade completely out of the wood. Gently push the blade into the surface of the wood until it's cutting all the way through. Then lock the blade at the proper depth and continue the cut.

4. **Finish the cut by hand if necessary.**

   Add some fasteners if needed to hold the plywood sheathing fast to the framing studs. Depending on how clean your cut is or how much extra wood may protrude into the opening here or there, you may need to shave off some excess with a handsaw or a powerful sander.

Save the piece of wood you just cut out! You may want to use it to create your coop's actual window or door panel, since you know it's the exact size you need. If not, it's always handy to have a few good-sized scraps of plywood lying around for miscellaneous coop add-on items or repairs.

Working with other materials

Of course, not every chicken owner wants a coop with plywood walls. Maybe you're interested in matching your own house to meet neighborhood covenants, using up some leftover siding from a home renovation, or just making your henhouse a bit more decorative for the sake of aesthetics. Here are a few of the more common alternatives to solid plywood or T1-11 walls, along with some things to consider for each.

Fiber-cement siding

Often known by a brand name like HardiePlank, fiber-cement siding is one of the most popular building products in the United States. Contractors love it because it's relatively easy to install, and homeowners appreciate its exceptional durability, rot resistance, and other low-maintenance qualities. It can be purchased in a number of different colors; some types even feature a faux wood-grain texture to mimic authentic wood siding.

It's installed in horizontal pieces that are generally about 12 feet long. The lengths are easily cut with a circular saw, but a carbide-tipped blade made specifically for cutting fiber-cement board is recommended. The bottom of each board overlaps the top of the board under it, allowing water to shed off the wall's exterior. (This overlap also covers up the nailheads for a clean, fastener-free look.) Some sort of weather barrier (such as the commonly-seen Tyvek housewrap or even roofing felt) is required behind the siding, sometimes with a plywood or OSB sheathing layer behind that. Small pieces of weather barrier are also placed behind butt joints where two pieces of siding meet end-to-end. This prevents water from seeping into the joints and getting trapped behind the siding.

TIP

Fiber-cement siding is nailed into place, either manually with a hammer or, more expeditiously, with a pneumatic nailer. See Chapter 3 for help deciding which tool and method are right for you, but be aware that many manufacturers recommend drilling pilot holes before driving siding nails with a hammer, a task that adds considerable time to the job.

When you're using fiber-cement siding, the trim goes on first! Pieces are ripped to size and fastened to corners, with the siding itself being cut to size to fit snugly in between. (See Figure 8-2 for a visual.)

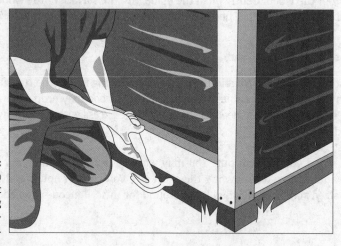

FIGURE 8-2:
Install the trim pieces first when working with fiber-cement siding.

REMEMBER

It's imperative that the bottom row — the first board to be attached to the coop — is perfectly level. Use a guideline (marked with a level or chalk line, as described in Chapter 5) to help ensure that all subsequent rows are straight. A strip of scrap wood or a short ripped siding board at the very bottom of the wall is completely hidden by the first full piece of siding, but ensures that the bottom board "kicks out" the same way that the others do.

Recommended tools, fasteners, and techniques for installing fiber-cement siding differ from brand to brand. Most manufacturers print and distribute their own detailed instructions on how to install their product. Be sure to grab whatever literature you can from the home supply center and consult the manufacturer's Web site to help you achieve perfect results.

Vinyl siding

Few DIYers purchase vinyl siding specifically for their coop project; some may have a stack left over from a previous project and be looking for a way to use this "free" and no-maintenance material on their coop.

Vinyl siding needs plywood or OSB sheathing behind it. A layer of insulation can be sandwiched between the sheathing and siding, but many DIYers don't bother with this step for a utilitarian structure like a coop. Those who do find that large sheets of rigid foam (nailed in place with special nails that feature big plastic washers) or a thin membrane of housewrap (stapled to the plywood) are usually the simplest solution.

Vinyl siding is cut most easily with a good pair of tin snips (see Chapter 3 for more on this handy tool), but a fine-toothed plywood saw blade or even a sharp utility knife will do the job, too.

The correct trim is vital to a quality vinyl siding job. Special J-channel trim pieces and under-sill pieces can help you achieve a clean look. And in the world of vinyl, trim pieces are usually installed first, before the siding itself.

Install the siding from the bottom up, with each 12-foot length locking into the piece below it. All nailing is done along the top edge of each piece, in specially designed nailing slots.

Vinyl siding should never be nailed completely fast against a wall! Vinyl siding will expand and contract with temperature changes, so the panels themselves need to allow for that movement. A gap of 1/32 inch should be left between each nailhead and the siding, or just enough that the piece of siding slides back and forth when pulled.

Board-and-batten siding

Unlike fiber-cement and vinyl siding, which are oriented horizontally across a wall, board-and-batten siding runs vertically. It is made up of wide *boards* (usually 1x12s or 1x10s) butted up to each other side by side. Narrower *battens* (often 1x3s or 1x2s) run parallel and cover each seam. The look is well-suited to a ranch or Western-themed coop.

Because board-and-batten siding runs vertically, horizontal blocking must be installed first. This blocking, often 1x4 furring strips fastened across and on top of the wall studs, provides necessary extra nailing surfaces for the boards. (Figure 8-3 shows what's happening behind the boards.)

Because board-and-batten siding is made up of basic lumber, you use basic tools to work with it. Cut the pieces to length with a circular saw or miter saw. Use exterior-grade nails compatible with the type of lumber you're using to fasten each piece to the furring strips. Use a long level to repeatedly check the plumb of each piece as you work to keep everything straight.

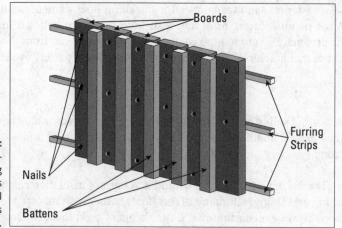

FIGURE 8-3:
Board-and-batten siding requires horizontal furring strips to nail into.

Boards

Furring Strips

Nails

Battens

Building a Basic Shed Door

Whether you're constructing a full-size walk-in coop or a small, chickens-only hideaway, a door you can open and close and lock shut at night is a key piece of the chicken coop puzzle. Even in the smallest of coops (like the A-frame coop in Chapter 13), a hatch is like a door, allowing the caretaker to access the shelter's interior for cleaning and bird-catching purposes.

But a prehung door (a self-contained unit with a door already hanging in a frame) is overkill for most coop builders. For starters, prehung doors will only fit in a walk-in coop; they just don't come in chicken-only sizes. And second, they can be a real bear to install, even for an accomplished DIYer.

Instead, you can make a very simple shed door using pieces you already have on hand and techniques you've already mastered by this stage of the build. And, it's far cheaper than buying a prehung unit!

Although it's typically called a *shed door*, the door described in this section works well in a walk-in coop or can also be scaled down to a very small size to create a chickens-only door in a small shelter. Here are the basic steps to building your own shed-style door. See Figure 8-4 for an illustrated look.

1. **Set aside your door panel.**

   If you sheathed your coop in plywood and cut out the door opening as described earlier in this chapter, you're done with this step already. It's easy to use that cutout as the door itself, but if you trashed that piece, cut a new one from a solid piece of plywood that's the same thickness as the exterior walls and the same size as the doorway opening.

   TIP

   Many people use a series of vertically standing boards to build a door panel. Two-by lumber stock generally results in a door that's too heavy, but 1x6s or 1x8s often make for an attractive and sturdy door. Lay the boards side by side (you may have to rip a board to get the correct door width); they'll all be fastened to the bracing pieces you add in Steps 2 and 3 to hold the door together as one solid unit.

2. **Cut framing pieces for the interior.**

Using 2x4 lumber, make a frame for the interior perimeter of the door panel. Create a rectangular frame that's ½ inch shorter and ½ inch narrower than the door panel itself. This provides the necessary clearance for the door to close without binding. Add one more horizontal piece across the middle of the door, between the vertical slats of the rectangular frame.

TIP

You may also choose to install *slam strips* on the inside of the door's frame. These extra pieces of framing, added as trim, are what the door actually closes against. Scrap wood works fine for this purpose; fasten it to the doorway's frame using nails or screws every few inches. Position the slam strips inside the door's frame by the thickness of the door panel itself, so contact is made when the door is closed.

3. **Add "Z" pieces.**

You now have two rectangles framed out, one atop the other. For each rectangle, position a 2x4 so that it runs diagonally from one corner of the door's bracing frame to the opposite corner, like the letter Z (see Figure 8-4a). Use a pencil and a level or straightedge to transfer the horizontal edges of the top and bottom boards to the diagonal 2x4, and then cut along the lines. These 2x4s should now fit snugly between the top, middle, and side braces on each rectangle, and will give the door additional rigidity. (On small chicken-only doors, these diagonal "Z" supports typically aren't necessary.)

4. **Fasten all pieces to the door panel's interior.**

Be sure each piece fits snugly against the others as you drive your fasteners. (Gaps or loose-fitting boards will only get worse the more the door is used.) Use nails or screws that won't go all the way through the receiving piece of lumber. If your door panel is made up of 1x6s, use two fasteners per board; 1x8 door panels may require three.

5. **Attach trim pieces to the door's exterior and doorway frame.**

This is a matter of preference and style, but doors look more finished with trim pieces around the perimeter of the door's exterior. Also consider adding exterior trim pieces where the hinges will go to give the hinge's screws more wood to grab onto. And you'll probably want trim outlining the opening of the doorway, too, to totally finish the look.

Whereas the door's interior trim pieces are cut to be slightly shorter than the door panel itself (to allow it to close), exterior trim usually butts right up along the door panel's edges for a more finished appearance.

Exterior trim typically serves more of a decorative function than a structural one, so you can go easy on the nails or screws. A pair of small trim nails or screws every 18 inches or so is usually sufficient for attaching these pieces to the door panel.

6. **Add door hinges.**

T-hinges (as shown in Figure 8-4b) or similarly-styled strap hinges are best for attaching a door to the coop wall. Three will suffice (two for a small, nonhuman door). Opt for a bigger, heavier hinge than you think you need. (A small hinge may weaken under the weight of the door and heavy use over time, causing the door to sag.)

Fasten the long part of the hinge to the door first with the screws included, making sure that they thread into blocking pieces, not just the thin door panel. Then have a helper assist you in raising the door into position while you fasten the other side. Fasten the short side of the hinge to the coop wall so that the screws penetrate solid 2x4 framing behind the wall sheathing.

7. Add some sort of latch or locking mechanism.

You can attach a doorknob to the completed door, but a simple hook-and-eye, sliding barrel bolt, or hasp is often sufficient; you're not trying to thwart professional cat burglars with a sophisticated Fort Knox-like security system. Keep it easy enough that you can latch or unlatch it while carrying a bucket of water, an armful of fresh eggs, or a wriggling chicken. But don't make it *too* easy; the rule of thumb says that if a 5-year-old can open it, so can a raccoon.

REMEMBER

Catches, latches, and similar pieces of hardware are meant to be adapted by the user to fit his or her specific situation or needs. Follow the instructions that come with your hardware and use the fasteners provided whenever possible. But be aware that some coop designs will require you to get creative with the hardware, the fasteners, or perhaps both. You may have to swap the hardware's screws for shorter or longer ones, or nuts and bolts to make a secure attachment that lasts.

WARNING

For a small chicken door that you won't use yourself for coop entry, a self-locking gate latch may be super-handy. But it's probably not what you want on a walk-in coop. Think about it: you go in one day to do some cleaning, and a stiff breeze blows that door shut, locking you inside your coop. A latch that can be undone only from the outside doesn't sound so smart now, does it?

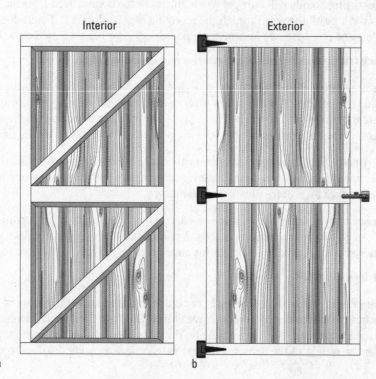

Interior   Exterior

FIGURE 8-4:
A typical built-from-scratch shed door.   a   b

Making Your Own Window

A window simply makes your chicken coop a nicer place for your flock (and for you, when you're out there working in it!). Windows serve several important functions on almost any coop design:

>> First and foremost, they permit sunlight to enter the coop.

>> If the window opens, it admits fresh air and aids ventilation.

>> Chickens love to sit inside the protected shelter of their housing and enjoy a view of what's going on out in the world, just like you do.

On smaller coops, the window may actually be a door. If there isn't enough wall space to justify a door *and* a window, the hatch-like door that leads outside can also perform the same functions as a dedicated window.

As with doors, it's pretty easy to make a decent window for a larger chicken coop out of miscellaneous parts and pieces. It's not like your birds need all the bells and whistles that come with double-hung units, like tilt-open sashes and UV-blocking films that protect furniture from fading. (Many manufacturers, though, produce small, no-frills, prehung units for tool sheds and similar structures. These can usually be purchased or ordered at big-box retailers or building supply centers.) Many coop builders have had success with the following build-it-yourself method, seen in Figure 8-5:

1. **Create a solid window panel just like a door panel.**

    Using Steps 1 through 5 from the earlier section "Building a Basic Shed Door," cut a plywood panel to fit the window's framed opening.

2. **Add hardware that allows the window to be opened and latched.**

    Add T-hinges at the top of the panel so that you can lift it up from outside the coop (follow the guidelines in Step 6 in the earlier section "Building a Basic Shed Door"). A hook-and-eye latch can be added to hold it open, or a simple prop stick can do the trick.

3. **Add wire mesh inside the coop.**

    Use fencing staples or screws and large fender washers every few inches to securely fasten heavy wire mesh inside the coop, covering the entire window opening. Voilá: instant window screen. (Chapter 10 deals with fastening wire mesh in more detail.) Open the window for sunlight and fresh air, and even leave it open in summer without fear of a nighttime raccoon raid. Or close it up tight to keep the shelter warm and dark.

If you're lucky enough to live where temperatures are always mild (between 40 and 90 degrees), you may be able to leave your window open all the time! Lots of coop builders have secured a wire mesh screen over an opening in the wall and then called it a day. If the window is well-protected from rain and you're not worried about ever completely closing up the coop to maintain winter warmth or insulate against intense summer heat, your window can be a permanent vent.

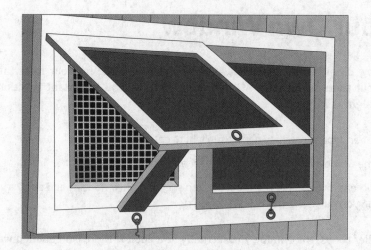

REMEMBER

Building a coop is largely an exercise in improvisation. Many backyard chicken owners decide to wing it with many aspects of a shelter, and windows are a popular place to get crafty. Got an old window that was salvaged from a home renovation project? If it's still in one piece, it can probably be retrofitted with a little blocking here and some trim there to work in your coop. Many DIYers have gotten by just fine with sheets of clear acrylic, Plexiglas, or Lexan, either screwed into place over a wall opening or mounted to a simple hinged frame built from scratch.

Topping Off Your Coop with a Roof

If you're constructing your coop from the ground up, the roof represents the final (and perhaps most important) big step of the build. The roof will keep your flock dry, which is, of course, the main function of any shelter. Certain coops, like the A-frame we provide plans for in Chapter 13, are really nothing *but* a roof, omitting walls from the design altogether!

While a few enterprising owners we know have gone all *Survivor* and done some impressive things with a heavy tarp, most DIY coop-builders have two options, really, when it comes to a roof: asphalt shingles or corrugated panels. (Both of these materials are explored more in Chapter 4.)

Hanging out your shingles

The shingle roof, as seen on millions of homes everywhere, is a multilayered weather and moisture barrier that has remained virtually unchanged in its basic makeup for decades. Shingling a roof isn't a job that many weekend warriors attempt because it's usually time-consuming, dirty, unpleasant, and (by its very nature atop most houses) particularly danger-ous. That's why roofing a home is almost always left to a crew of professionals.

But shingling a chicken coop brings the job down considerably in scale. A smaller roof means a lot less square footage to cover, and even the largest chicken coop won't put you so high off the ground as to be a life-or-death hazard (although climbing a ladder always requires some extra care). And the fact that a coop is usually a simple, free-standing structure eliminates many of the tricky aspects of regular roofing, like how to keep leaks from occurring in a valley where two downward-sloping roof pieces meet.

With nothing more complicated than one peak to deal with, shingling a coop roof is a DIY-friendly job. Here are the basic steps for shingling:

1. **Plan and put down sheathing.**

    Your framework of roof rafters needs a solid layer upon which you'll install the outermost cladding. Roof sheathing is almost always plywood (OSB is also commonly used) and is generally between ⅜-inch and ⅝-inch thick.

    Measure your roof and cut plywood or OSB to size. It's best (and will be stronger) if the sheathing's long edge — and thus the wood's grain — sits perpendicular to the roof rafters below. On extremely large coops, you may need more than a full sheet to cover each section of roof. Panel ends should always rest on rafter centers, so measure and cut your sheets of plywood accordingly.

    To secure the sheathing, start at the bottom edge of the roof, nailing or screwing the plywood or OSB to the rafters below. Follow standard plywood-nailing techniques as described in the earlier section "Fastening plywood in place" — use 6-inch spacing around the edges, switch to 12-inch spacing in the middle, and don't countersink your nails.

TIP

You can install *fascia boards* from the roof for a finished look; these boards hang vertically from the rafter tails, sitting parallel to, but out away from, the walls. Be sure to factor in the thickness of the fascia boards with the roof sheathing. For example, if your fascia boards are ½-inch thick, let the roof sheathing hang ½ inch over the ends of the rafters. (Use an actual piece of the ½-inch board as a spacer on each rafter tail to make sure things line up before you nail.)

REMEMBER

Plywood panels need a thin gap between them to allow for expansion. On a roof, the 4-foot edge should be resting on a rafter. Along the 8-foot edges, use one H-clip (see Figure 8-6) between each pair of rafters. These small metal clips help maintain a smooth, flat surface from one piece of plywood to another and keep the expansion gap intact.

Work toward the peak of the roof, staggering subsequent rows of sheathing (like bricks in a wall) so you never have a vertically-running joint longer than 4 feet. Use half-sheets of plywood if you must in order to accomplish this. (Check out Figure 8-7 for a visual aid.)

TIP

Check your progress with a long level often. Use *shims* (thin, tapered wedges of wood, sold at any hardware store, that can be inserted between two surfaces during building to fill a gap) underneath a rafter here or there if needed to get a level surface. If the sheathing isn't perfectly flat, any waves or humps will transfer right through to the shingles and be painfully obvious. Your chickens probably won't care, but it looks bad and will drive you nuts every time you see it.

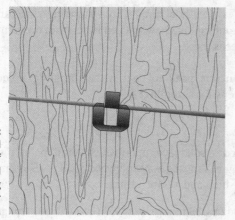

FIGURE 8-6:
H-clips hold sheets of plywood sheathing together on a roof.

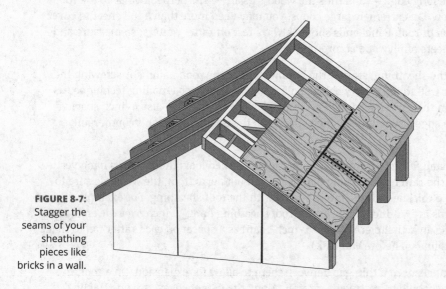

FIGURE 8-7:
Stagger the seams of your sheathing pieces like bricks in a wall.

2. **Install fascia boards, drip edge, and roofing underlayment.**

Builders often rip their fascia boards out of 1x6 lumber (or something similar) for a perfect fit, as the angle of the rafters can sometimes make for rafter tails with odd measurements that stock lumber won't always cover neatly. (But as long as the rafter tail is covered, some extra overhanging material won't bother your birds.)

Measure the total length of your roofline, from outside rafter tail to outside rafter tail, and cut a fascia board to match. Holding it level (you may need assistance for this), nail it twice into each rafter tail.

After attaching your fascia boards to the rafter tails, install *drip edge* (as seen in Figure 8-8). This metal flashing protects the fascia boards and helps water shed safely off the roof. Lay it flat against the fascia and nail it securely. Use one roofing nail in each rafter tail.

Next, a waterproof barrier must be sandwiched between the shingles and the roof's sheathing. Use rolls of heavy-duty tar paper or roofing felt. Roll it out, starting at the bottom, cut it to length, and staple or tack it down to the sheathing, one fastener per 12 inches. With the next row, overlap the first by 4 inches. Work your way up to the peak.

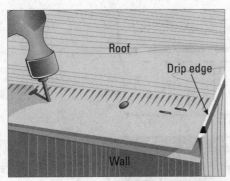

FIGURE 8-8:
Drip edge is an
inexpensive
way to help
protect your
roof from
water.

Use the lines marked into the tar paper to keep the rows straight; it'll help you keep the shingles themselves straight later.

TIP

3. **Start with a starter strip of shingles.**

Most asphalt shingles have three tabs. The two splits that create these tabs allow water to hit whatever's directly below. So for the lowest row of shingles, you need a layer of shingles that no one will ever see to protect the sheathing underneath. This first row of hidden shingles is called a *starter strip*, and it's really important to a good roofing job.

Begin by snapping a chalk line on the tar or felt paper you installed in Step 2 to use as a guide (see Chapter 5 for tips on using a chalk line). The line should be $11\frac{1}{2}$ inches off the bottom edge of the roof if you're using standard 12-inch shingles. This allows $\frac{1}{2}$ inch of the bottom row of shingles to hang over the drip edge for rain runoff.

Only use blue-colored chalk for marking lines on a roof, because red chalk can stain some roofing materials!

WARNING

With a straight guideline clearly marked, cut 6 inches off the end of one shingle with a sharp utility blade (see Figure 8-9a). Now flip it upside down so the tabs point up to the peak of the roof (see Figure 8-9b). Align the shingle with your chalk line and with the edge of the roof, and fasten the shingle to the sheathing with roofing nails — one near each end, and another just below each slot between tabs. Repeat this for the rest of the starter strip. Trim if necessary at the other end of the roof.

Some shingle manufacturers sell special starter strip shingles designed just for this purpose. Use them if you think they'll make your life considerably easier, but there's nothing wrong with the tried-and-true method of cutting your own as described in this step.

TIP

4. **Continue shingling.**

The first row of shingles goes right over the starter strip, this time with the triple-tabs pointing downward, the way you're used to seeing them. That 6 inches you cut off the edge of the first starter shingle now allows you to butt a full shingle against the roof edge without the slots lining up. Each new course of shingles should cover the darkened portion of the shingle below it so that each long shingle now looks like just three narrower tabs. On a typical 12-inch shingle, this translates to 5 inches of visible shingle sticking out, called a *reveal*.

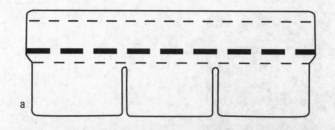

a

b

Nail through two shingle layers (never just one), and be sure all heads are hidden by the shingle on top of it. You want one nail near each end, and another just above each slot. Trim 6 inches off the first shingle in every other row to maintain staggered tab slots.

REMEMBER

After every four courses or so, make sure you're working in a straight line with your shingles. Measure from the bottom edge of a shingle up to the nearest line marked into the tar or felt paper you installed in Step 2. Repeat this at several spots along the same row, making sure you get the same measurement every time. If you're a little bit off, just start readjusting the next row of shingles to get yourself back to level.

5. **Cap off the peak with cut shingles.**

At the peak, you can't leave those nailheads exposed on the topmost rows of shingles. Flip a shingle over and cut it into three sections, as shown in Figure 8-10a. Be sure to cut upward from each tab slot at an angle, to get three tabs that are narrower at the top than the bottom. (Toss the two scrap triangles away.)

Now start at one end of the peak, bending each tab over the peak, and nailing it in place (see Figure 8-10b). Continue with cut shingles until you reach the middle of the roof, and repeat the procedure at the other end of the peak until your cut shingles meet. Cut one more shingle, using a 5-inch chunk of the finished shingle color. Nail this *closure cap* over the joint in all four corners (see Figure 8-10c).

TIP

If you want to put in a ridge vent at the peak of your coop roof, you can omit this row of cut shingles because you'll just end up tearing them off. See "Venting Your Coop" later in this chapter for more on this option.

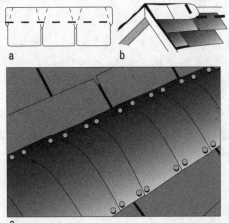

FIGURE 8-10:
To cover a roof peak, cut shingles as shown.

a   b   c

Conquering corrugated panels

Not everyone who wants a roof over their chickens' heads relishes the idea of a long, dirty, monotonous day working with traditional asphalt shingles. If you're searching for a shingle shortcut, consider corrugated panels.

As we discuss in Chapter 4, these panels can be heavy plastic, fiberglass, or even metal. They feature a series of grooves and channels to help with water runoff. Their large size (usually around 2 feet wide, often in lengths of either 8 or 12 feet) generally means you'll need to hang just a couple for your coop, instead of a couple dozen (or couple hundred) individual shingles!

TIP

Fiberglass roofing panels are usually translucent, meaning that they admit a good deal of light! These panels can go a long way in letting sunlight into the shelter; you may not need to include a window in your coop plans.

WARNING

The downside to fiberglass panels is that they're more susceptible to some types of damage than a traditional shingled roof. If your coop will be located in an area where heavy limbs routinely fall from older trees, know that a fiberglass roof provides little protection. In such a location, you may be better off with shingle-and-sheathing roof construction, an entirely new coop location, or calling a professional tree-trimmer.

To install corrugated roofing panels, follow the manufacturer's instructions for your particular panels. In general, however, follow these guidelines:

1. **Use proper support pieces under the panels.**

    Most panels require support pieces that run perpendicular to the panels. These *closure strips* (see Figure 8-11) feature a contour to match the panels. This contour provides proper support under the entire panel, whereas simple straight boards support only the bottoms of the channels, leaving the tops of the ridges unprotected and more suscep-tible to breakage. Secure closure strips atop the rafters at regular intervals.

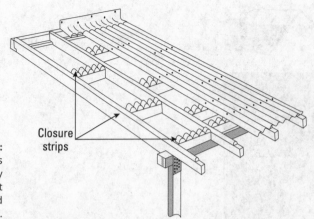

FIGURE 8-11: Closure strips properly support corrugated roofing panels.

2. **Start at the bottom.**

   Work your way up toward the highest point on the roof, overlapping new panels on top to allow water to shed safely over the seams.

3. **Use proper fasteners.**

   Most panels need to be predrilled before they're secured with fasteners. Most also require a weather-tight seal at each fastener hole, in the form of either a rubber gasket or a dollop of silicone sealant. The fasteners are generally placed on the high sides of the panels' raised ridges to keep the fastener out of a channel full of rainwater.

Venting Your Coop

Good ventilation is essential to any coop of any size, for the health of your flock as well as your own comfort when working in the coop. (We go into more detail about ventilation in Chapter 2.) Give some thought to how you'll air out your coop as you plan it so that you can properly build in adequate ventilation for your chickens. Ventilation can be added after the coop is built on an as-needed basis, but it's obviously easier to incorporate it up front.

REMEMBER

Any method of venting requires not just a way for stale air to escape, but also a way for fresh air to enter the structure. Normally, your coop's door and/or window will be opened often enough throughout the course of a day to allow this to happen on its own. Just another good reason why heading out to the coop on a regular basis is part of a chicken-keeper's job!

Here are a few popular ventilation methods (all shown in Figure 8-12).

>> **Wall vent:** The easiest method is to cut a hole out of one or more exterior coop walls and add a basic vent screen, similar to what you would find in every room of your home (see Figure 8-12a). Decorative wood models are also available. Wire mesh can be used as well.

REMEMBER

A vent should be placed high on the wall, since the ammonia-soaked air you're trying to exhaust will rise inside the coop.

» **Ridge vents:** Ridge vents are built into the shingled roof of a structure. These raised pieces cap the roof's peak, but feature narrow gaps underneath the peak itself through which air can escape (see Figure 8-12b). They're often purchased along with roof shingles so that the style and color match, although plain metal versions exist that can be fitted to any roof. Some ridge vents require you to cut a gap in the roof's plywood sheathing to allow them to operate properly.

WARNING

» **Cupola:** For a real down-on-the-farm look, consider a *cupola*. These small structures added onto a shingled roof are a time-tested and architecturally attractive way to allow air to exhaust to the outside (see one in Figure 8-12c). Installing one can up the degree of difficulty on the roof construction phase of your coop build considerably, though, so proceed with caution if you're a carpentry novice. (Or have a handyman on stand-by to assist with this part of the project.)

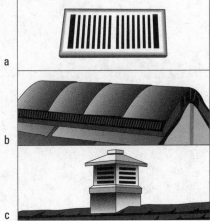

FIGURE 8-12: Wall vents, ridge vents, and cupolas all allow you to vent your coop.

a

b

c

Chapter 9

Building Creature Comforts

Have you ever walked through an empty house for sale? It's a little cold and sterile with all that cavernous space and no furniture. Sure, you can imagine where your sofa, bed, and that vintage '70s bean bag chair will go. But without furnishings, it's just a house — not really a home.

The same holds true for your chicken coop. Yes, you have a sturdy floor and good walls all around. And that roof overhead may keep out the rain, but it doesn't automatically qualify your newly-built structure as a *chicken coop*. Heck, even after you throw your hens in there, it *still* could be considered nothing more than a toolshed that a few chickens happened to wander into.

Arguably, it's not a true chicken coop without a few key pieces that are, oddly enough, the easiest pieces of all to build. This chapter deals with the two basic pieces of "furniture" that every coop needs: a roost and a nest box. As a bonus, it addresses a third piece that's necessary for some coops to keep you from leaving your flock high and dry: a ramp.

Finalizing Your Flooring

In Chapter 4, we mention how some chicken owners put down a layer of linoleum or vinyl flooring over the plywood or OSB floor to aid in coop cleaning. If that sounds like you, now's the time to install it, before you start fastening things like roosts and nest boxes to the floor. These features will be more difficult to work around later.

Whatever material you'll be using, cut it to the size of the coop. Working on a freshly-swept, clean floor, lay the material down to test the fit, and make any necessary adjustments by trimming the material.

If you're putting down linoleum or vinyl flooring, use appropriate adhesive. Or use wide-headed roofing nails instead! (You'd *never* do this inside your home, but as we keep saying, your chickens won't care in the least. It'll be covered with loose bedding material anyway!)

For a wire mesh floor, use fencing staples every few inches to secure the material to the plywood subfloor. Get the wire mesh tight against the floor to lessen the chances of a bird getting her toes caught under a loose section.

Coming Home to (a) Roost

Your coop's *roost* is where the chickens will sleep at night or just hang out when they're chillin' in the coop. Chickens are especially vulnerable to attack when they sleep, so they prefer to do so as high off the ground as they can get. In the wild, they'd find a low-hanging tree branch or fence rail to camp out on, and more than likely, return to the same spot each night. Building a simple roost into your backyard coop offers your birds that same nightly protection. Even if you have free rangers or leave the coop-to-run door open full-time, your chickens will almost certainly head inside once the sun goes down and take their usual spots on their roost for the night.

In this section, we explain how to select your roost's location, describe options for making a roost, and tell you how to safely secure it in your coop.

Location, location, location

In a smaller coop (like the minimal coop in Chapter 12), there simply aren't very many places to put a roost. But for a larger coop where you have multiple options (like the all-in-one coop in Chapter 15), several factors can help determine your coop's most righteous piece of roost real estate.

Size

REMEMBER

Allow each bird about 12 inches of roost. For a modest flock of 6 chickens, that means you need at least 6 feet of roost. "But wait a minute," you might be saying. "I don't want to build a coop that's 6 feet wide! And what if I want to get more chickens? Do I need to build an add-on wing just to extend the roost?!" Fear not. Breaking up the roost into several smaller roost sections is common practice. (More on this "stair-step" approach in the sections to come.) As long as they're at least 2 feet apart (to give your birds plenty of room to stretch and flap while roosting), you can install as many small roosts as your flock needs to give each hen 12 inches of roost to call her own.

Height

Out in the wild, chickens will roost as high as they can for maximum protection from things that go chomp in the night. Your coop's roost should be located as high as you can install it.

Make it at least 2 feet off the floor if space allows. Lighter birds can go higher — 5 feet or more — *within reason.*

TIP

Think about sleeping on the top bunk of a bunk bed. Yes, you can lie down and *sleep,* but sitting straight up in bed can lead to a nasty headache. Make sure you allow enough room for each bird to sit fully upright on the roost.

Drop zone

Perhaps the most important consideration when planning a roost's location is what's right underneath the roost. No matter what your coop layout looks like to you, whatever's directly below the roost bar is always the bathroom as far as your flock is concerned.

REMEMBER

Never put the roost right over feed and water dishes, and if a nest box is to be located under the roost, be sure it has a cover. (We discuss nest boxes in detail later in this chapter.) Also, you may decide that right over your doorway isn't the best place for the roost, unless you plan on wearing a rain slicker every time you enter the coop.

TIP

Coop-builders who need to break up a roost into smaller sections may be tempted to save on space by stacking one roost directly over another. Great idea — unless you're on the bottom roost. No one wants to wake up in the morning covered in chicken poop — not even a chicken! If you need to stack your roosts, spread them out in a "stepladder" orientation. This is easy for the birds to climb and gives each of them a "clear shot" from each roost to the floor below. (An example is shown in Figure 9-1.)

FIGURE 9-1: A roost laid out like a stepladder gives all your chickens their own safe space.

No matter where you place your roost, the drop zone is going to need some cleaning. Here are some ideas for making cleanup easy:

>> It's common for caretakers to place a "poop board" under the roost to catch droppings and help keep the coop floor less messy. The board is either removed for cleaning or scraped clean as needed.

>> Some coop-owners build a part of the shelter without a solid floor, placing the roost over the hole, which they then cover with mesh wire. The droppings (or most of them) fall

through the wire mesh into a pit, where they can be dealt with outside the coop with a shovel and wheelbarrow. Be aware that this method could make your coop much colder in winter months due to the big gaping hole in the floor.

>> Many coop-owners have gone a long way in minimizing their clean-up time by installing dropping pans underneath their chickens' roosts. A plastic tub or open-top kitty litter box works well for this purpose (as shown in Figure 9-2a). Some crafty DIYers have taken it a step further by rigging up a large tray (see Figure 9-2b) that can be slid out of the coop through an exterior hatch. Upon removal, the tray is cleaned and then replaced for no-mess manure cleanup.

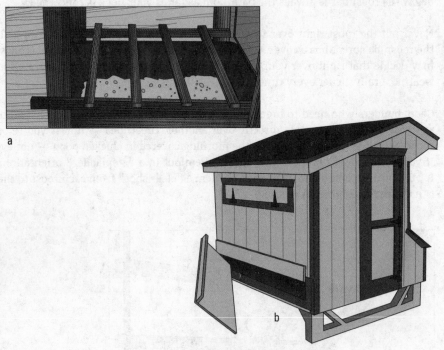

FIGURE 9-2: A droppings pan (a) or removable tray (b) can help keep a coop clean.

Roost requirements: Making your roost out of different materials

While some thought needs to be given to where you put your chickens' roost, what you make it out of matters a lot less. The following are popular options:

>> **Lumber:** Roosts are often just long pieces of scrap wood. (Most of our Part 3 plans use a 2x3 leftover from the coop framing.)

There's some debate among flock folks about how much lumber a chicken actually prefers for its perch. Some argue that the bird likes to be able to grip the roost with its feet, wrapping its toes around the narrow board. Think of a 2x4 standing on its narrow edge. Others feel that the chicken would rather sit flat-footed on a broad roost and turn that 2x4 over or use an even wider board. It's hard to give one blanket rule that applies to all chickens, because preferences can also vary by breed. Be prepared to swap out your original roost for a new material if your birds seem uncomfortable using it.

Chicken owners in particularly cold climates may want to consider using a wide roost for a very practical reason. A chicken sitting flat-footed on a wide board tends to cover its feet with its belly, and the belly feathers keep the delicate toes warm during bitterly cold nights. If frostbite is a genuine concern for you and your flock, wider is better.

>> **A dowel:** Some owners prefer the rounded shape of a dowel to the squared-off edges of a 2x4, because it more closely mimics the tree branches that birds use in the wild. (Chickens we asked did not state a preference one way or the other.) Wooden dowels are easily found at hardware stores and come in a range of lengths and diameters.

Generally speaking, don't use a dowel that's less than 1½ inches round (and even that's a little small). Anything less is probably too flimsy to support your flock. It's not worth traumatizing your chickens if their roost breaks in two during a peaceful night's slumber. Think "closet rod," not "broomstick."

>> **A tree branch:** Some über-crafty types go so far as to recycle a fallen tree limb from the yard! Cut to the proper length and mounted securely inside the coop, a heavy branch *really* replicates the outdoor experience for your birds and gives your coop interior a natural and rustic feel.

Securing and supporting your roost

You have several options for securing and supporting your roost inside the coop. Many caretakers opt for a freestanding roost rack, a series of roost bars that are affixed to a self-supported, framed structure that can even be moved around from spot to spot inside the coop. (An example can be seen in our Walk-In coop in Chapter 16.) Others use the stepladder approach described earlier and shown in Figure 9-1. These "stepladder roosts" are usually fastened in place using nails or screws through the stepladders' vertical legs: at the bottom into the coop floor, and at the top into the coop ceiling or wall. (See Chapter 5 if you need to review proper toe-nailing technique.)

For just a simple roost bar suspended between two coop walls, one nail or screw in each end is often sufficient. You'll be toe-nailing through the roost bar into the wall of the coop, but check the length of your fastener to ensure that it doesn't go all the way through the wall and leave a nasty sharp point sticking out the exterior of the coop. (You could also drive your nail or screw from outside the coop, going through the exterior wall and *into* the end of the roost bar, but finding precisely the right spot on the outside wall so that your fastener lines up perfectly with the roost bar is harder than it sounds.)

A final option is to make *nailing cleats:* small, scrap pieces of lumber that are securely mounted to the wall with short fasteners. Then the roost bar is positioned on top of the cleats and fastened in place by nailing or screwing straight down through the roost bar and into the cleat. This method generally gets you out of having to do any toe-nailing whatsoever, because all connections are made at easy 90-degree angles.

Whatever technique you utilize to secure your roost, make sure the entire thing is solidly anchored for wobble-free mounting and dismounting. A rock-steady roost provides comfort and stability that your birds will appreciate.

Many chicken owners have discovered a way to make cleaning the area under the roost a little easier: hinges! Attaching one end of the roost bar to the coop wall with a hinge allows the entire roost to be lifted out of the way when it's time to clean the coop. (Scrap wood can be used to build a "notch" to accept the roost bar on the opposite wall and still keep it stable when in use.) It's a $2 fix that will pay for itself every time you don't have to maneuver a long-handled broom or shovel over, under, and around a series of horizontal 2x4s!

REMEMBER

Depending on the material you've used for your roost, the length of the roost bar, and the number and size of the chickens who will be using it, you may need to add some support underneath the roost. It shouldn't bow or sag, even if every chicken in your flock is on it at once. Prop it up if needed with a length of 2x4 fastened to both the roost and the coop floor; just drive a fastener through the roost bar into the top end of the 2x4, and toe-nail the base end of the 2x4 into the coop floor.

Feathering a Nest Box

If you got into keeping a backyard flock so that you could enjoy "free" fresh eggs whenever you want them, then perhaps no part of the coop is more essential to you than the nest box. These small cubbies provide your birds a plush and protected place where they can deposit their eggs and where you can collect them at your convenience. And while some finicky hens may never decide to use them for laying, a chicken coop isn't really a chicken coop without a few nest boxes tucked safely inside.

In this section, we give you pointers on designing, placing, and building nest boxes; we also offer options if you prefer not to build a box from scratch.

Designing nest boxes

As you design and build your coop, give careful thought to your nest boxes. Check out the following sections for a few considerations to factor in.

Size

A nest box typically needs to be 12-x-12-inches square. That's a bare minimum; bigger is usually better for most breeds. The idea is to give one hen enough room inside the box to turn around completely. If your nest box has a cover, make it tall enough for your hen to fully stand up inside the box.

WARNING

But don't build them too big! Oversized nest boxes can lead to hens sharing a stall, and that often leads to broken eggs. A nest box bigger than 16 x 16 inches is probably starting to ask for trouble.

Number

TIP

Don't compare nest boxes to beds, where every member of the household needs their own. One nest box for each hen is considered to be a colossal waste of precious coop space, and often results in some boxes that go completely unused. Instead, think of your coop's nest boxes like bathrooms in a home. No one spends too terribly long in them, so as long as there are enough to meet demand, everyone gladly shares the facilities. One nest box for every two to three hens is right on the money.

Placement

Nest boxes should always be placed together in a group, rather than spread around the coop in ones and twos. (Hens like the social aspect of laying eggs.) Figure 9-3 shows a group of side-by-side nest boxes built as one unit (you can also build several separate boxes and set them close to one another). Stacking nest boxes on top of one another is fine as long as you don't place any of them much higher than about 3 or 4 feet off the floor.

FIGURE 9-3: Side-by-side nest boxes let your birds be social.

Hens prefer their nest boxes to be in a dark, protected area of the coop. Directly across from a door or window is generally not the best placement. Because the elevated roost is usually located in the darkest corner of the shelter, many owners place their nest boxes underneath as a space-saving technique. (Just remember to put a full cover on any nest box underneath a roost to avoid contamination from chicken droppings!)

It's become increasingly popular for coop-builders to build a nest box that sits *outside* the coop, as shown in Figure 9-4. Attached to an exterior wall, the hens can come and go while inside the shelter, but a hinged top allows you to gather eggs without entering the coop or disturbing your hens. Several of our coops in Part 3 feature this type of nest box setup.

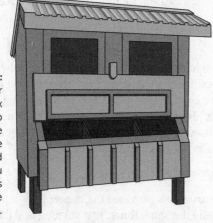

FIGURE 9-4: An exterior nest box frees up floor space inside the coop and gives you free access from outside the coop.

If you know you're going to incorporate your nest boxes into the exterior wall, the following tip may be helpful. While planning your stud layout during framing of the coop walls, leave the studs at the nest box location 25½ inches apart (for two boxes). This will allow you to easily slip the boxes in between later without having to cut out part of a stud.

TIP

On any exterior nest box, make that lid lockable or latchable, to thwart any four-legged eggnappers from helping themselves in the middle of the night. Also, install some sort of weather guard over the hinged lid seam to prevent rainwater from leaking into the nest boxes and making a soggy mess. A simple flap of rubber liner or heavy-duty plastic sheeting might suffice, or you can get creative with an inexpensive piece of aluminum flashing from the roofing department of your building supply center.

TIP

Many coop-builders with interior nest boxes cut a hinged access panel in the exterior wall that allows them to reach inside to gather eggs from outside the coop. (See the All-in-One coop in Chapter 15 for an example.)

Bedding

Remember that after you construct a nest box, you'll need to line it with some sort of bedding material to maximize your hens' comfort. Loose straw or soft hay is generally the ideal choice, although some caretakers have had success with wood shavings or even shredded paper! Keep it loose and about 3 or 4 inches thick so that your birds can get in there, kick it around, rearrange it, and fluff it up to their liking. (And don't forget to replenish it as needed; clean nesting material means clean eggs.)

Building nest boxes

After building an entire chicken coop, slapping together a few nest boxes will be an easy add-on project. In fact, you'll likely be able to do it using scrap pieces of plywood and leftover lumber! Follow these steps for each box:

1. **Build the skeleton.**

   Use short lengths of framing lumber to construct a frame that fits your coop's specs. Build a hollow cube using the same techniques as you would for framing the coop itself, as described in Chapter 7. The principle is the same: same lumber, same fasteners. Make sure the framing pieces don't eat up too much of the nest boxes' interior space.

2. **Clad all but one side in plywood.**

   Plywood is usually the material of choice here because you probably have some scrap pieces leftover from the coop build. Use small fasteners to secure the plywood to the lumber frame (just like cladding exterior coop walls — see Chapter 8), making sure that your sharp-pointed fasteners don't protrude into the open space of the nest box itself. This same sheet material can be used to create dividers between nest boxes in a multinest structure. Keep one side open as the birds' entrance/exit.

3. **On an exterior box, add a lid or hatch for easy access.**

   For a side-opening hatch, follow the instructions provided in Chapter 8 for building a door; a nest box access hatch is essentially the same thing, just smaller, and hinged on the top or bottom instead of the side.

If you like the idea of a top-opening lid, you'll probably want to make the nest box walls sloped so that a lid sitting on top won't allow rainwater to collect and drench your eggs (and maybe your birds) when you open the lid. Nest box walls that are taller at the coop wall and get shorter as they extend out allow the top lid to function like a hip roof and let water safely shed away from the hinged opening. Simply hinge the top plywood panel at the coop wall and add some sort of latch or locking closure to keep hungry predators from helping themselves.

REMEMBER

Be sure to construct your nest boxes so that they're all exactly identical to one another. Having one nest box that's wider or taller than the others, one that gets more or less light than the others, or even one that's painted a different color will almost certainly lead to fighting within your flock. Hens often pick one box as their favorite no matter what, and you may see some ruffled feathers from time to time over who gets to use it, but don't invite problems by making one nest box more or less attractive than the rest.

Other options: Buying or repurposing nest boxes

Some people prefer to purchase nest boxes made of metal or heavy plastic. (Periodic cleaning may be easier, but these are no better, really, than wooden DIY models.) Some can be mounted to your coop wall while others have handles and are meant to be portable. (These can come in handy if you have to temporarily relocate a hen outside the coop or if you've just increased the size of your flock but haven't had time to build additional nest boxes yet.)

Other chicken owners decide that, rather than build or buy a nest box, it makes sense to them to recycle some other item and turn it into a nest box. A tall storage tub no longer being used, with a hole cut in the front, makes a perfectly acceptable nest box. A covered kitty litter box is the ideal size and shape for a nest box. Even plastic milk crates can easily be transformed into nest boxes with just a little ingenuity.

Ramping Up

Most every coop has a small door for the chickens to pass in and out of the shelter. But many coops, because of their particular design, have that chicken door located well off the ground. Many rookie coop-builders realize only after they finish their structure that their flock is stranded inside with no safe way to get out into the run! (Imagine if, instead of using your front door, you had to leave the house every day by jumping out of a second-floor bedroom window. Getting back in would be tricky, too!)

The majority of chicken caretakers employ a basic ramp to help their birds get in and out of the coop. As shown in Figure 9-5, it's just a board that stretches from the door to the ground. It's a great way to use up a long piece of scrap from the coop build. (A fence board, usually about 1x8, is the perfect size and can be purchased in easy-to-work-with lengths of 5, 6, and 8 feet.)

TIP

There are no hard and fast rules, really, about how to design or construct a chicken ramp. But most chicken-folk have come to a few generalizations that seem to hold true in most circumstances:

>> **Adjust your angle:** The angle of the ramp will depend on how high off the ground the chicken door is located. A steeper angle, though, will obviously be much more difficult for your birds to navigate day in and day out. There's no magic number, but anything steeper than 45 degrees is less of a "ramp" and more of a "slide."

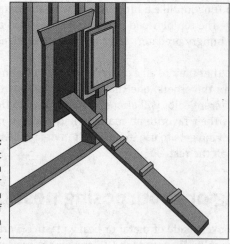

FIGURE 9-5:
A basic ramp can help your flock get in and out of the coop on its own.

>> **Rig up some rungs:** Most coop-builders add rungs to their ramp that run crossways (refer to Figure 9-5), giving the birds steps to grip with their toes. Again, almost any scrap pieces of lumber leftover from the build will do the trick. How far apart you space your rungs will depend on the angle of the ramp and your birds; you may find that you need to add more rungs if they're having a tough time. (Use screws instead of nails to fasten them; they'll be easier to pull up and reposition as needed.)

>> **Take the stairs instead:** If your ramp is unusually steep or your chickens just can't get the hang of it, you may be better off coming up with some sort of makeshift staircase for them. Some chicken-keepers firmly believe that the birds prefer a series of stairs to an angled ramp anyway. You could use stacks of cinderblocks or even build a simple set of "stairs" out of scrap wood, as shown in Figure 9-6.

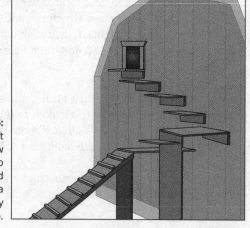

FIGURE 9-6:
Makeshift stairs allow chickens to come and go from a two-story coop.

WAITING FOR THE ELEVATOR: HELP FOR RAMP-CHALLENGED CHICKENS

Many chicken owners are constantly amazed and surprised at the cleverness and intelligence that their birds seem to display. But plenty of others realize that their flock isn't exactly made up of rocket scientists and Mensa members. Online forums like www.backyardchickens.com are littered with stories of people who built awesome coops with convenient and well-designed ramps, only to have their birds completely baffled by how to get into or out of the coop!

More than one caretaker has found himself going out at dusk to find a flock of birds huddled around the bottom of the ramp, patiently waiting to, one by one, be lifted up by hand and gently placed into the coop, where they presumably expect to be tucked in with a goodnight story. Then, in the morning, those same chickens stand around inside the coop until their ride to the run shows up to personally escort them back outside. (Hey, maybe these birds *are* smarter than we give them credit for, after all!)

People who have experienced this kind of behavior from their chickens are generally able to "teach" them to use the ramp by placing the birds on it and coaxing them up or down, either with feed or a little nudge. Often within just a few days, the chickens are scampering up and down the ramp all on their own.

Before you tear down that brand-new chicken ramp or completely redesign your just-finished coop to add a new ground-level entryway for your birds, know that you may just need to provide a little bit of coaching for your chickens to grasp the concept of the ramp.

Chapter **10**

Assembling a Run

he vast majority of this book has dealt with constructing the shelter portion of a chicken coop: the physical lumber-and-nails structure that serves as shelter and housing for your flock. But when most chicken owners talk about their "coop," they're also including the run: the exterior, usually-contained, often-open-air "yard" that's adjacent to the shelter.

Chickens need (and prefer) to spend a good deal of time outside in the sunshine and fresh air. With all this talk about building little houses for them, it's easy to forget sometimes that chickens are outdoor creatures; we erect small-scale homes for them solely to make raising chickens easier and more rewarding. So it follows that the run must be an important consideration when designing and building a backyard coop.

For many modest-sized coops, the run is indeed part of the structure itself — a small, fenced-in patch of bare ground that's covered overhead by the same roof as the indoor shelter. (Of the five coops we provide detailed plans for in Part 3, three of them feature this "under-one-roof" type of setup.)

Many backyard chicken-keepers, however, find that their flocks need more room to run. Perhaps they're creating a coop out of an old toolshed or playhouse that doesn't have a built-in pen, or they simply want to keep more birds than a small all-in-one coop can accommodate. Others simply have a lot of property and want to turn their chickens loose to work as much soil as possible.

While a select few are blessed with conditions that allow them to let their birds go free-range, most chicken owners need to keep their feathered friends protected from predators by containing them within the safe confines of a run. Should you need to construct your own chicken run separate from the coop, this chapter will come in handy.

Framing a Simple Run

A run should provide each bird with 3 to 6 square feet of yard, but more would certainly be appreciated by the chickens. For instance, say you have a flock of 10 hens — not a huge flock, by any means. Theoretically, you could get away with a run that's 30 square feet, but it would be better to build a run that's 60 square feet or larger!

TIP

If you're just getting into chickens, you may be tempted to opt for a smaller run, so it doesn't eat up your entire backyard and risk raising the ire of a spouse or neighbor who's not so sure about the whole bird business. But if the chicken bug bites you (it usually does) and inspires you to supersize your flock, you may find yourself limited by the puny pen you built to begin with. Assuming that space allows, try to build a bigger run than you think you need. This way, you can easily add to the flock later on without having to crowd your chickens or build a larger (or secondary) run.

The best run is attached to the coop itself with a small door in the coop that allows the birds to come and go freely between coop and run whenever their little chicken hearts desire. (An example can be seen in Figure 10-1a.) Alternatively, some large runs may be free-standing and completely surround the coop, offering the same carefree coop-to-run access for your flock (as shown in Figure 10-1b).

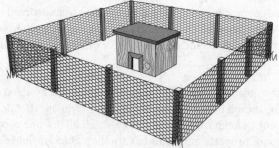
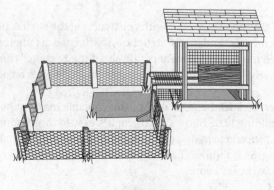

FIGURE 10-1: A chicken coop with an attached run (a) gives your birds an open-door policy, while a perimeter-style run (b) surrounds the whole structure.

While the golden rule of runs is 3 to 6 feet of space per bird, that deals only with square footage on the ground. There is no magic formula that specifies how *tall* your run should be. But you do need to think about it ahead of time, because the height of the run determines the size of your fence posts and the wire mesh you choose. Chickens aren't big flyers, so creating your own 20-foot-tall aviary exhibit is a waste of time and materials. But many coop owners like to make the ceiling of the run high enough to accommodate themselves (even if their coop is a tiny box); it's much easier to walk into the run to retrieve a hen than to crawl in and chase her around on your knees.

Here are the basic steps for framing an attached or free-standing run:

1. **Determine the run size and layout.**

   Assume 3–6 square feet per bird, and measure out the run on the ground. Here's where you can play with all the options. Try both an attached run and a freestanding perimeter run if you have room. Consider different shapes: a square, a rectangle, an L, a free-form blob. All the birds care about is having enough space; they don't care about the layout. Go with what looks best to you and what works best in your yard.

   Mark the outline with stakes, colored flags, or landscaper's marking paint. This allows you to step back and double-check the size and views. Adjustments are easy to make now; after you dig a few post holes, not so much.

2. **Set the first fence post.**

   If the run will meet the coop, start with the post closest to the structure. For a free-standing run, pick a corner. How you set the all-important first post depends on whether it's wooden or metal:

   - Wooden posts like 4x4s or corral poles (as mentioned in Chapter 4) need to have holes dug for them. Use a manual, clamshell-style post-hole digger or a power auger, as explained in Chapter 6. You may choose to permanently secure this first post with concrete or simply brace it in an upright position so you can dig all your holes at once.

   - If you're installing metal T-posts, as discussed in Chapter 4, use a post driver to sink the first one directly in the ground. (You can encase metal posts in concrete if you want the added security, but most people who choose them do so to get out of messing with big holes and wet concrete.) Check the above-ground height of the post with your fencing material to make sure you don't bury the post too deep.

   No matter what type of post you use, check for plumb with a level or post level (see Chapter 3) before moving on.

3. **Set the post at the far corner and fill in between.**

   Sinking one post after another in a row can sometimes lead to an uneven line, unless you're super-meticulous. An easy way to keep all your posts in a straight row is to make an opposite corner post the next one you sink, so you have one post at each end of what will be a long fenceline.

   Once this second post is secure, attach a long masonry string to the face of each post. Orient this length of string near the top of the posts and add a second string just above the ground. The bottom string tells you where to dig your holes or sink your posts. The top string helps you position the posts themselves in a straight line and keep them all standing plumb, as shown in Figure 10-2.

How far apart should your posts be? That depends on your fencing material. Assuming you're using a good, heavy-gauge, welded wire (as suggested in Chapter 4), the spacing may be dictated by the width of the fencing panels or the roll of wire. If you're stretching the roll out in one long piece, you'll need posts every 4 to 6 feet or so to maintain proper support for the fencing. The sturdier the wire, the farther apart the posts can be. Flimsier fencing will require more posts set closer to one another.

4. **Repeat Step 3 to create the other walls of your run.**

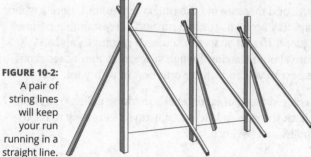

FIGURE 10-2:
A pair of
string lines
will keep
your run
running in a
straight line.

TIP

While you may be able to dig all your holes at exactly the correct depth and end up with all your posts sitting at just the right height, you're more likely to need to *top* your posts once they're all in the ground and totally secure. This usually involves running a level string line along the tops of the posts, marking them at the appropriate spots, and trimming the tops with a saw as they stand in the ground. See Chapter 6 for more on how to do this safely and accurately.

Working with Wire Mesh

With a series of posts set, all that's left is to stretch your fencing material around the posts and fasten it down. Most runs incorporate some type of rolled wire mesh. Your material of choice is determined by how strong you want the wire mesh to be. Look for strong, heavy-gauge wire with openings small enough that a predator can't reach in and inflict any damage to an unsuspecting hen. A short rundown of commonly-used fencing materials is found in Chapter 4.

REMEMBER

As discussed in Chapter 4, "chicken wire" would seem to be the ideal material for a chicken run, but it's far too flimsy, with openings that are far too big, to be of much defense against the majority of predators. It's not recommended for constructing a run.

Sizing up, measuring, and cutting the wire

Most backyard coop-builders opt for a fencing material that is sold on a large roll. Depending on the gauge of the wire and the store you're looking in, that roll can often be found between 2 and 5 feet tall in lengths as short as 25 feet and as long as 100 feet. What that means is, for the vast majority of coop projects, you'll end up buying *a lot* more fencing than you need. But having a stockpile of this versatile material is usually a good thing.

A heavy wire mesh product has all sorts of practical uses — both chicken-related and not — for the average caretaker. Better to buy a bigger roll than you need of the good stuff and have extra than downgrade to a flimsy chicken wire simply because the hardware store sells it by the foot. The first time a hungry predator forces its way through your run and makes off with a bird or two, you'll probably regret your choice.

Measuring and cutting wire off a humongous roll can be tricky. Most DIYers find it easiest to roll the fencing out on a big patch of bare ground, be it the backyard or the driveway. Weigh down the free end with cinder blocks or a couple of 5-gallon buckets full of chicken feed, and carefully walk the roll backward, unrolling as you go. When you've stretched out what seems like an adequate amount, measure it with a tape measure (walking on the material as you work won't hurt it and can help you get an accurate measurement) and use a pair of tin snips (see Chapter 3) or other wire cutters to cut the wire to the desired size.

REMEMBER

When measuring and cutting wire fencing, remember to allow a few extra inches of material so that you can secure it to your end posts. For example, a 10-foot run of fencing may mean 10 feet from post to post. Cutting your wire mesh right at 10 feet may not leave much room for making the attachments. Ten feet plus several inches on either end gives you plenty to work with (and ensures even more mesh-to-post contact, which means more places to fasten it down and a sturdier run overall).

WARNING

Be sure to wear heavy work gloves for this step, as most wire mesh types tend to leave extremely sharp pointed ends behind as you work your way through a cut.

WARNING

Some owners select rigid panels of welded wire. The set widths of these panels dictate how far apart your posts need to be set, because the panels can be extremely strong and difficult to cut, even with tin snips. To trim down a panel, you may need to break out a hacksaw or a heavy pair of bolt cutters.

Fastening the wire to your posts

To secure your wire mesh to the wooden posts of your run, you have the following fastener options:

>> **Fencing staples:** Also called *poultry staples,* these U-shaped nails (see Figure 10-3) are specifically designed to fasten wire mesh products to wooden posts. They're easy to find (in the nail aisle of any hardware store), inexpensive to buy in bulk, and when they're installed properly, they're surprisingly sturdy and strong (as anyone who's tried to remove them from an old fence post can attest). But putting them in can be extremely tricky — and painful. The staples are small — much shorter than nails. Holding them with your fingertips while you hammer them home is often darn near futile. (In fact, the bandage manufacturers should partner up with the poultry staple folks and sell combo packs.)

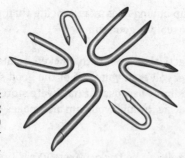

TIP

Holding the staples with a pair of needle-nose pliers can save your fingers, but it's not a foolproof technique. The staples are notoriously hard to grip (no matter how tightly you squeeze) and notoriously easy to bend (no matter how carefully you hammer).

>> **Screws and washers:** A method we particularly like uses ordinary screws. Driven into place with a drill and a screw bit, they provide a lot more holding power than staples and stand up better to the prying attempts of a hungry raccoon. But the heads of most screws slip right through the gaps in most wire mesh products. That's where large washers come in handy.

Use a large *fender washer* to hold the fencing down, and then drive a screw through the washer's hole (as shown in Figure 10-4). Now the washer is what's really securing the fencing, with the screw holding the washer in place.

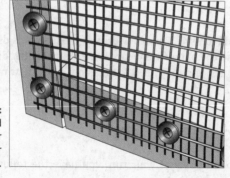

FIGURE 10-4:
Screws and
large-diameter
fender
washers.

REMEMBER

If you're using metal T-posts on your run, you shouldn't need fasteners. The posts have small tabs on them that clamp around the wire mesh. But some coop-owners have found that particularly crafty predators can force the fencing out of these tabs. Consider using metal wire ties to help hold your mesh in place on metal posts. (Nylon zip ties may be handy, but can be chewed through or even broken by a determined critter.)

Don't be skimpy with the fasteners you're using to secure your fencing material. Unlike nailing together some 2x4s, just a few fasteners in the corners aren't going to cut it. A determined predator will seek out the weak spot in your run fence and use it as his attack point. You should plan on placing a fastener every few inches on every single post. Then, when you're done, give the fence a few hard tugs, adding more fasteners if the wire mesh gives at all.

Adding even more wire

TIP

In addition to installing fencing along the sides of a run, many coop-builders also run some sort of fencing material over the entire top of the run to combat flying predators. If you want to use the same fencing material overhead, some additional roof truss-like pieces of lumber may be required to help support the mesh. Many people, though, have good success with plastic netting (found in most garden centers) or even with lengths of string or twine strung across the run in a makeshift canopy. Lightweight materials like this may not stop a dive-bombing hawk who's really determined (or a predator who can climb and jump), but they should stall an airborne attack long enough for your hens to run for cover.

TIP

It's also a good idea to extend your run's fencing material down into the ground and actually bury it! You might think that most predators would see your well-constructed run and just give up. But imagine that same predator with an empty belly and eight or more hours' worth of darkness on his hands (or paws). That's plenty of time to do a little digging, meticulously burrow under the fencing, and snack on some fresh chicken nuggets. This, of course, requires a little extra quality time with your shovel, but it will be time well spent. Just dig a trench 12 to 18 inches deep between all the fence posts. Then attach the fencing material to the posts as normal, securely fastening it to the posts both above the ground and down in the trench. Finally, backfill the trench to bury the fence.

Chapter **11**

Plugged In: Basic Electricity for Your Coop

The following is a true story (or it easily could be). Only the names have been changed to protect the innocent.

John was a hard-working guy who decided one day that he wanted his family to raise their own chickens. He checked local codes and obtained the proper permits. He researched different types of chicken coops. He thoughtfully considered and meticulously gathered the various tools and materials he'd need. He blocked out several weekends, took his time, and built a spectacular little coop that he was rightfully quite proud of. Then, at a grand ribbon-cutting ceremony with his wife Mary and their children in attendance, six beautiful hens were welcomed into their new home.

One Saturday morning not long after, Mary woke up early with plans to cook a big family breakfast with fresh, just-laid eggs from the coop. She put on a pair of slippers and a housecoat, grabbed a basket, and plodded out into the pre-dawn darkness.

Minutes later, John was awakened by the slam of a door and a stifled, guttural yell. He raced downstairs to find Mary, her slippers covered with foul-smelling chicken poop. Another long slick of it striped up her backside. The viscous, slimy yolk of several broken eggs coated her hands and ran down the front of her housecoat. And just above her steely, glaring eyes, perfectly centered on her forehead, was a huge golf-ball-sized welt, already turning swirling shades of black and blue.

"You couldn't put a *<%$@#^!! light out there?!?" she snarled.

Despite the lovely and expensive flashlight he gave her for Christmas later that year, John found himself solely responsible for his backyard flock and all related chores from that day forward. Especially collecting eggs in the dark.

Running electricity to your chicken coop may seem like an over-the-top luxury at first glance. But it isn't just about providing adequate lighting for you during early morning egg runs and maintenance visits; chickens need lots of light to facilitate egg production. A coop with power is also much easier to heat in the winter, cool during summer months, and keep properly ventilated year-round. And let's face it, those things will ultimately make your coop a more pleasant place for you and your chickens.

This chapter deals with getting electricity to your coop and how to best use it for lighting, heating, and ventilation. If only John had read it. . . .

Calling a Pro to Run a Line to Your Coop

REMEMBER

Unless you happen to be a professional electrical worker or are infinitely confident in your wiring skills, it's probably best to leave this phase of the job up to a licensed electrician. Tapping into existing power lines, working with exterior-grade electrical conduit, and adding a sub-panel to your home's circuit breaker box is not the kind of project most DIYers are comfortable attempting. It's frighteningly easy to set something on fire or send yourself to the hospital (or worse) by getting in over your head with a tangle of hot electrical wires.

Leave this up to a pro: Call a licensed electrician and describe your chicken coop project. For someone who works with this stuff every day, running power to a backyard coop is a fairly simple and straightforward job that should be able to be completed within a few hours. Once a working line has been run to the coop, it's a snap for you to add specific fixtures or components if you want. Of course, an electrical professional will be more than happy to take on the entire job, too: putting outlets, switches, and light fixtures anywhere and everywhere you specify. If you're at all nervous about working with wiring, don't hesitate to farm out this part of the build to someone who does it for a living.

WARNING

You may be saying to yourself, "Why can't I just throw a long extension cord out there?" It's a reasonable question: you probably have an outdoor outlet to spare, and a heavy-duty extension cord will likely cost less than an electrician's service call. So why mess with anything more permanent? Like so many other things, it's about safety. Outdoor extension cords are meant to be temporary helpers in running juice where you need it: for power tools during a building project, to run that leaf blower for a few hours over the weekend, or to light up your yard with a holiday display. Nonstop use will quickly overload and burn out a cord. And having a thin, 120-volt extension cord stretched out along the ground where your chickens are constantly pecking with their sharp beaks is just a recipe for disaster. (Fried chicken is great, but not like this.)

Buzzing About Outlets versus Switches

Once a qualified electrician runs a line to your chicken coop and collects his check, you'll likely be left with a single cable made up of 3 thinner color-coated wires inside. The several inches of cable that the contractor left for you should be curled up in a weatherproof junction box that's been secured to a wall stud or ceiling rafter. Power is coming from the house's main panel to these wires. Now it's up to you to decide what to do with that power. In general, you have two options: a receptacle (more commonly known as an outlet) or a switch.

Receptacles (also known as outlets)

A receptacle is what we nonelectricians typically refer to as an "outlet." (But if you want some instant street cred in the electrical aisle of your local hardware store, whip out some contractor slang and shorten it to "recep.") It's what we plug something into to make it work. And by and large, it's your best and easiest choice for power in a chicken coop.

Most coops can get by just fine with one duplex receptacle like you have in your house, with slots for two devices, one above the other. For all but the biggest coop, this should provide sufficient power. (If you need more on a temporary basis, a multi-outlet power strip often fits the bill. And if you choose to add one more permanent receptacle, tying it in to the professionally installed one is usually a snap.) Wiring a single receptacle is as basic as electrical work gets. Just have your electrician install his junction box in the exact spot where you want your outlet to go, and you're already halfway home!

REMEMBER

As with lumber, sheet goods, and so many other materials used in a coop build, electrical receptacles, conduit, and junction boxes come in indoor-use and outdoor-use varieties. Exterior-grade components will have extra insulation, gaskets, and other safeguards to prevent rain, snow, or the spray from an overzealous garden hose from causing a short-circuit. Outdoor receptacles, for example, have built-in ground fault circuit interrupters (GFCIs) that must be used in any location where wetness could be an issue. For maximum safety, use only outdoor-approved supplies, even if they'll be located inside the structure of the coop.

Here's how to install an outdoor-approved GFCI receptacle (see Figure 11-1):

1. **Cut all power to the area where you'll be working.**

REMEMBER

   Electrical work always carries the potential for real danger, so this first step should be considered nonnegotiable. Always shut off the power at the breaker box before beginning any wiring. Double-check to make sure you threw the correct switch with an inexpensive circuit tester. (Never assume that the labels on your breaker box switches are correct, even if a professional electrician has been there before you.) You may even want to put a piece of tape over the breaker switch to prevent a "helpful" family member from flipping it back on while you're working.

2. **Strip the ends of the electrical wires.**

   Using wire strippers, cut ½ to ¾ inch of colored insulation away from the ends of all wires. This includes the black (hot), the white (neutral), and the green (ground) wire, as shown in Figure 11-1a. (Some ground wires are bare, uninsulated copper that need not be stripped.)

3. Form a hook in each wire and secure it to the receptacle.

With needle-nose pliers, bend the bare end of each wire until it forms a hook. This hook will be looped around a screw terminal on the side of the receptacle. Each terminal is labeled to help you match each wire to its appropriate terminal.

Loop each wire around its screw terminal so that the wire runs clockwise. This way, the hook closes around the screw as you tighten it down. (If you loop the wire counterclockwise around the terminal, it can spread itself open as you tighten, possibly resulting in a failed or unsafe connection.)

WARNING

Most receptacles also have an alternative method of attaching the wires via a series of holes on the back of the unit. To *backwire* a receptacle, do not form hooks in the stripped wires. Instead, simply insert each bare wire end straight into the appropriate hole until the internal locking mechanism clicks into place and "grabs" the wire. But a word of caution: Backwiring is considered to be a less reliable method of wiring a receptacle and a technique that many professional electricians neither employ nor fully trust.

4. Screw the receptacle to the junction box and add an appropriate cover.

Use the screws on the top and bottom of the receptacle to secure it to the corresponding screw holes in the junction box. Then be sure to add an exterior-grade outlet cover, as shown in Figure 11-1b. When the outlet is not in use, this cover will also keep the receptacle safely closed off from wayward beaks, wings, and feet.

5. Test the GFCI to make sure it's working properly.

Plug in a radio or lamp to the GFCI and restore power to the area by flipping the switch at the breaker box. You should be able to turn on the device you plugged in. While the device is on, press the button on the receptacle marked "test." If the device shuts down and the receptacle's "reset" button pops out, everything is perfect. Push "reset" to restore power to the GFCI. (Refer to Figure 11-1a to see the locations of both buttons.)

If the "reset" button doesn't pop, something is wrong. Kill the power and retrace your steps, double-checking all wire connections to make sure that they're tight and that all wires are fed to the appropriate terminals. If you're still having trouble after one do-over, it may be time to seek the advice of a pro to get you back up and running.

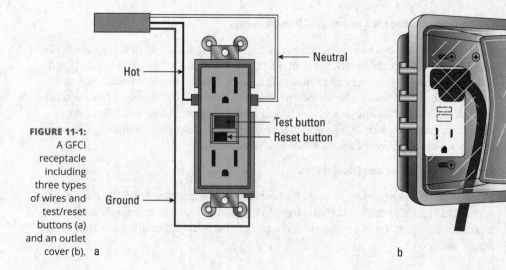

FIGURE 11-1: A GFCI receptacle including three types of wires and test/reset buttons (a) and an outlet cover (b).

REMEMBER

If you head out to your coop one day to find no power, make the GFCI the first item you check. Sometimes a simple surge can trip the receptacle's onboard circuit breaker and shut the device down as designed. Many an hour has been wasted troubleshooting everything from light fixtures to electrical wires and connections to circuit breaker panels . . . only to find that a simple push of the "reset" button would have brought the juice back on-line in an instant.

Switches

A switch controls electrical power running to another fixture. Installing one involves more materials and more work than wiring an outlet. You can easily hook up a switch to the wires left for you by the electrician, but you'll have to run more wires off of the switch to the thing that the switch will control, and also install another junction box to house *that* electrical connection.

Then there's this to consider: Say you decide to wire a handy light switch right inside your coop door. You wire it up correctly and run more wire up to a ceiling-mounted junction box, from which you hang a very cool light fixture. That's awesome, until you want to plug in a heater. Or until you realize that you need more than just one bulb in the coop. Or you want to add a fan for ventilation purposes. Now another light bulb goes off over your head: this one, the imaginary cartoon version that comes with the sudden realization that you should have just put in an outlet or two instead.

REMEMBER

In the rest of this chapter, we examine some plug-in lighting options that work just as well as ones you control with a switch. Because the installation and use of receptacles is easier, cheaper, and more versatile in the long run than switches, that's what we recommend for most DIY chicken-coop-builders.

Letting There Be Light

All chickens need varying degrees and amounts of light just to live normal chicken lives. (They won't even eat in the dark!) It's essential that you provide your flock with a source of natural light; the outdoor enclosure of a run (as described in Chapter 10) and a shelter built to include windows (see Chapter 8) fit the bill nicely. But some chicken-owners find that supplemental artificial light in a coop is also a necessity. This section sheds some light on why a light bulb or two may be a good idea, and it explains where and how to include some lighting in your coop.

Deciding whether to include artificial light in your coop

Whether artificial light is truly necessary in a coop is a subject of some debate among backyard chicken owners. Many prefer to go *au naturale*, leaving all the lighting to Mother Nature. If you're aiming for big-time egg production, though, your chickens need 14-16 hours of bright light every day. Some owners raising meat birds provide full light 24 hours a day to encourage round-the-clock feeding! Many others argue that these practices are unnatural and even unhealthy for the birds.

Decide for yourself what your ultimate goal is: maximum egg production, raising birds for meat, or simply providing a home to a small backyard flock. Talk to other owners about what works for them and what they might do differently. You may find that adding supplemental lighting is an unnecessary expense and not worth the hassle and worry.

Ultimately, artificial lighting may be a "nice-to-have" item for you, but it's certainly not a "need-to-have" item for your birds. Chickens survived just fine for centuries before Thomas Edison invented the light bulb. Know that yours are capable of living full and healthy lives without ever seeing one for themselves.

Properly placing the right amount of lighting in your coop

The amount of light you need in your coop depends on its size. A good guideline is 100 watts of light per 200 square feet of shelter space. Since many coops are much smaller than that, it's easy to get away with very little supplemental light. If there's enough light for you to read, it's plenty of light for your flock.

WARNING

Don't overdo it! A light bulb has been called "fire in a bottle." Even a low-wattage bulb can generate a lot of heat. Putting one inside a small box made of wood makes many people nervous. Sadly, coop fires have started because of a too-powerful bulb burning for too long in a coop filled with wood shavings. When in doubt, go with less light for safety's sake.

Some coop owners find that it's best to have two light sources:

>> A main bulb is for dark days, early mornings, and late evenings to ensure 14–16 hours of bright light per day.

>> A smaller, dimmer bulb can be used as a nightlight all night long. Chickens tend to sleep extremely deeply and can become easily disoriented in total darkness. If you're worried about one of your chickens needing to find its way back onto its roost at midnight or defend itself against a nocturnal predator, a nightlight will provide peace of mind.

For the brighter main bulb, a centrally-located spot on the coop ceiling is best. This provides a maximum amount of light to a maximum percentage of the coop. Ideally, your chickens should not be able to reach the bulbs themselves. If they can, be sure to use an approved cover or wire guard to avoid broken bulbs inside the shelter.

The dimmer bulb should be placed where its soft, low-level light can just barely reach the roosting area.

Choosing fixtures

Your chickens don't care about the fixture that provides the light nearly as much as they care about the light that it provides. This means you can (and should) keep it simple. Think utilitarian all the way. Here are a few basic and inexpensive lighting options to consider:

>> **A clamp light:** A clamp light (see Figure 11-2a) is a basic light socket with a metal, dome-shaped shield that acts as a shade. These lights plug in to a receptacle via a cord of several feet, are turned on by a simple push button, and can be mounted to most surfaces thanks to a spring-loaded clamp. Some even come with a wire bulb guard. Want to adjust the lamp's location? Squeeze the clamp, move the fixture, re-clamp, and tilt or swivel the shade to direct the light exactly where you need it. This versatility will be appreciated as you reconfigure your coop interior from time to time, add more chickens, or do a thorough clean-out. A clamp light can also accommodate a red heat lamp in cold weather, doing double-duty as a heater (more on heaters in the next section).

>> **A trouble light:** A favorite of DIYers everywhere, a trouble light (see Figure 11-2b) generally has a longer cord than a clamp light, extending its reach significantly. It, too, plugs into the wall, features an easy push-button switch, and has a hinged cage to protect the bulb. Instead of a clamp, a trouble light has a hanging hook atop the fixture. Most trouble lights also feature a convenient electrical outlet built in to the handle. Fluorescent-bulb models exist as well.

>> **A shop light:** A hanging, 4-foot-long shop light (they also come in shorter and longer lengths) throws off a considerable amount of light, so it may be too much for anything but a very large coop. These plug-in fixtures (see Figure 11-2c) often feature two fluorescent bulbs and are usually operated by a pull chain or string. Some have built-in grids (to shield the bulbs from harm) and even onboard outlets. Some are not rated for outdoor use, though, so choose carefully.

FIGURE 11-2:
Good coop lighting options include a clamp light (a), a trouble light (b), and a shop light (c).

a   b   c

WARNING

DIYers who need a portable light source often choose the halogen work light. These rectangular fixtures are often paired on a tripod and use super-bright halogen bulbs to throw hundreds of watts of light around a jobsite. They also put off an incredible — even dangerous — amount of heat, and some are quite easy to tip over. Halogens should not be used for the purpose of lighting a coop.

TIP

Just because your chickens have specific lighting requirements before the crack of dawn and into the wee hours doesn't mean you have to wear a path to and from the coop to turn the lights on and off. If ever there was a perfect use for the common light timer, this is it. A simple timer, available at any hardware store, will plug into your receptacle (just another reason why receptacles are better than switches in a coop, as we explain earlier in this chapter) and run your lights for you. It's a cinch to set the timers to control both the bright bulb and the nightlight at various times of day, and just as easy to adjust those times to complement the sun's seasonal cycles.

Warming Up to Heaters

Chickens do best with temperatures between 40 and 85 degrees Fahrenheit. Although they're quite adaptable to both hot and cold climates, egg production takes a severe hit when temps stay above 90 or below 32 for long periods of time. And northern chicken owners know that during a cold snap where the mercury nears zero, frostbite starts to claim hens' toes and combs!

But do you need to provide an artificial heat source? As with lighting, many owners say no. Even in areas prone to freezing, if a coop is properly-built and relatively protected from wind, the chickens generally provide all the body heat they need to keep the coop above 32 degrees. That may sound bitterly cold to you, but it's as warm as your coop needs to be during winter months.

REMEMBER

Because the goal is simply to keep the temperature inside the coop a few degrees above freezing, you certainly don't have to go nuts with industrial-grade kerosene blowers or propane-fueled heaters. In fact, any heater that uses a flame is probably overkill, because it carries an added risk of fire and carbon monoxide poisoning if improperly vented. For the small space of a coop, a basic heat lamp or compact, electric plug-in heater is likely more than you'll ever need, even in the most frigid of climates. You're just trying to take the edge off, not turn the coop into a sauna.

Mind the placement of any heat-generating device you place inside a coop. A heat lamp bulb can be dangerously hot to the touch and should never be located low to the ground, where a clumsy bird could get too close. Even a small electric heater that pumps out gentle, warming heat should be placed away from the heavily-trafficked areas of the coop, just to be on the safe side. Finding the right (and safest) spot for a heater inside your flock's shelter will likely require some experimentation on your part.

TIP

An inexpensive thermometer hung inside the coop can tell you at a glance when the interior temperature is too cold. And many electric heaters have built-in thermostats to help regulate the amount of soft, gentle heat they pump out.

A BRIGHT IDEA THAT WARMS THE HEART (AND THE FLOCK)

In the confines of a very small coop, you may find that the regular bulb you use to light your coop also provides enough heat to eliminate the need for a heater altogether. Skeptical that this could be an effective heating method? Here's a true story: In the early 1970s, one chicken enthusiast made a trip north to pay a wintertime visit to friends. His friends treated him to a Christmas-morning breakfast of fresh eggs. When he inquired as to how this was possible in the dead of winter, they showed him their backyard shed, which doubled as a chicken coop. Every square inch of the shed's interior had been lined with tar paper, which acted as insulation and allowed a single light bulb to keep the entire shelter cozy and warm for the hard-working hens.

Ready for the kicker? This story took place in the remote community of Manigotagan, *over 100 miles north of Winnipeg!* So if it works in the frigid wilderness of Manitoba, it can almost assuredly work in your neck of the woods.

Falling for Fans

While a fan inside your coop can provide your chickens with a dog-day breeze, most coop-owners need not worry about artificially cooling the inside of the structure. As long as the temperature inside the coop generally stays below 85 degrees Fahrenheit, you shouldn't have any problems. The shade inside the coop itself and an open window are generally enough to keep a shelter adequately cool for a flock.

REMEMBER

If you do utilize a fan, keep it well off the ground — the closer to the ceiling, the better. Make sure there's a protective cage around the blades to keep your hens safe (and you, too, for when you absent-mindedly walk in there to do some maintenance). Finally, make sure that the fan isn't too powerful; your chickens should have room inside the coop to get out of the path of rushing air. And make sure it isn't pointed at the roosts, which would make sleeping uncomfortable.

A fan can also be put to good use as a tool to assist with ventilation in your coop. It doesn't take long for even a few hens to saturate the air with ammonia, which can be harmful to lungs — both yours and your chickens'. Here's an easy rule of thumb: If you're comfortable breathing inside the shelter, then your ventilation is fine. If there's a noticeable smell of ammonia or you find moisture buildup on the walls or ceilings, it's time to air the place out.

Proper ventilation also heads off nasty problems like mildew and mold, which can invite bacteria and disease and eventually lead to sick chicks. An open window or door is a good start, and some coop-building methods allow air to vent on its own (you can check out some of these simple techniques in Chapter 8). You may find, though, that adding a simple exhaust fan is necessary to keep the air inside the coop fresh.

3

Checking Out Coop Plans

Here it is: the really fun stuff. These five chapters show you how to use the information in Part 1 and the skills from Part 2 to build five different chicken coops, each one having its advantages.

Each chapter includes a detailed list of materials you need, an illustrated cut list that shows how to cut all your lumber, and assembly instructions to help you put all the pieces together.

Want a starter coop that's minimal in every way? Go to Chapter 12.

For a space-saving A-frame design, see Chapter 13.

Chapter 14 features a tractor coop that you can easily move around your yard.

In Chapter 15, we spotlight an "all-in-one" design that incorporates the best elements of several coop styles.

And for large flocks, Chapter 16 offers a 64-square-foot, walk-in coop that can house up to 30 birds!

Chapter **12**
The Minimal Coop

Maybe after reading Parts 1 and 2, you still have doubts about your own carpentry skills and are skeptical that you can build your own chicken coop. Or maybe you're just looking for the rock-bottom minimum in terms of how much material you'll have to

Handwritten margin notes (left and top):

RUN
3-6 sqft
18 × 36 ft
1 ft circle

Coop
3 ft fl
2-3 12 × 21

Door
2-12 when plen

buy, time you'll have to devote, and effort you'll have to exert to construct a shelter for your flock. We've got your back with a bare-bones, zero-frills design that still lets you truthfully say that you built your own chicken coop. (We won't tell anyone how easy it was if you want to let the world think you're a master carpenter.)

This coop uses a few unorthodox techniques that go against some of the standards we lay out in earlier chapters (like no inner studs on the walls). We've also designed it so that there isn't a single angled cut anywhere on the coop, and no ripping of any boards is required. (In fact, many lumberyards will make one or two free cuts on a full sheet of plywood or OSB *(oriented strand board)*, so you can even get out of making the long cuts yourself if you use our cut list to direct the sales associate who mans the big saw.)

Furthermore, all the measurements are in inches or half-inches — no quarters, eighths, or sixteenths. In addition, you'll have almost no lumber leftover if you follow our directions. All of this was done in keeping with the spirit of a "minimal" coop that requires only an absolute minimum of materials, tools, skills, time, and cuts, and leaves behind a minimum of waste. And it's a pretty darn decent chicken coop to boot.

Vital Stats

- **» Size:** True to its name, the Minimal is small — just a 4-x-4-x-4-foot cube.

- **» Capacity:** With 16 square feet of floorspace, the Minimal can actually accommodate a medium-sized flock of four or five birds.

- **» Access:** One entire exterior wall of the coop is hinged to serve as your access door. To maintain the minimal theme, the coop was designed without a dedicated chicken door.

- **» Nest boxes:** A triple nest box is located inside the shelter and is made entirely from the main build's scrap lumber.

- **» Run:** The Minimal does not have an incorporated run. See Chapter 10 for thoughts on how to design and construct your own.

- **» Degree of difficulty:** All straight-line cuts are made on either inch or half-inch marks. There are no inner wall studs, no roofing shingles, and no doors or windows to frame. This is as easy as it can possibly get, and it shouldn't take more than a single afternoon to complete.

- **» Cost:** We estimate the total cost of our materials list at just $200, meaning that this coop is also minimal when it comes to the price tag.

Materials List

This list represents everything we used to construct the Minimal. Feel free to change up some materials to suit your needs or to take advantage of materials you have on-hand. (See Chapter 4 for more on selecting materials.)

Lumber		Hardware		Fasteners	
1	4' x 8' sheet of $\frac{5}{8}$" OSB	4	concrete pier blocks	100	12d nails (or 3" wood or deck screws)
2	4' x 8' sheets of $\frac{5}{8}$" exterior-grade plywood		Corrugated metal roof panels (enough to create a roof 48" wide x 64" long)	120	7d nails (or 1¼" screws)
2	8' lengths of 4x4 pressure-treated lumber	3	5" strap hinges	40	1" fencing staples
13	8' lengths of 2x4 lumber	1	chain lock	20	fasteners for corrugated roof panels
		12	galvanized rafter ties		
		2	12" x 48" pieces of wire mesh		

Cut List

Here's how to cut your lumber to create each piece of the Minimal coop. The individual pieces are then put together according to the illustrations. (For more about making safe and accurate saw cuts, refer to Chapter 5.)

Shelter floor

Quantity	Material	Dimensions
1	Sheet of $\frac{5}{8}$" OSB	48" x 48"
2	2x4s	48" long
4	2x4s	45" long

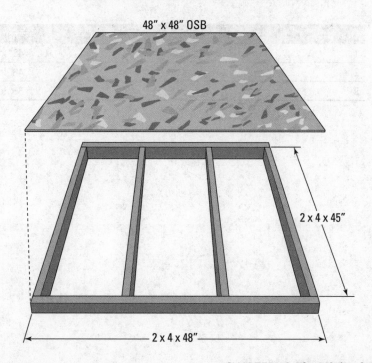

48" x 48" OSB

2 x 4 x 45"

2 x 4 x 48"

Front wall

Quantity	Material	Dimensions
1	Sheet of plywood	48" x 48"
2	2x4s	48" long
2	4x4s	45" long

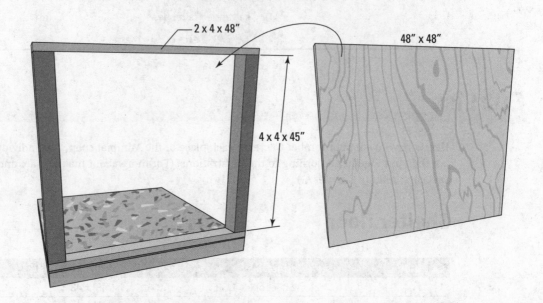

2 x 4 x 48"

4 x 4 x 45"

48" x 48"

Back wall

Quantity	Material	Dimensions
1	Sheet of plywood	48" x 36"
2	2x4s	48" long
2	4x4s	33" long

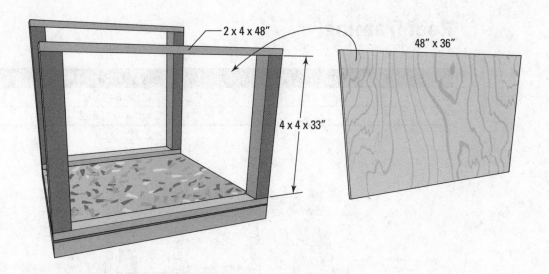

2 x 4 x 48"

48" x 36"

4 x 4 x 33"

Right and left walls

Quantity	Material	Dimensions
2	Sheets of plywood	48" x 36"
4	2x4s	41" long
4	2x4s	33" long

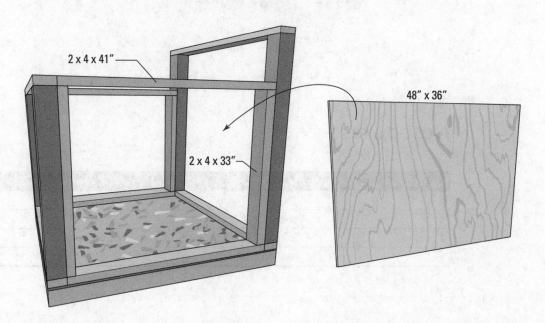

2 x 4 x 41"

2 x 4 x 33"

48" x 36"

Roof framing

Quantity	Material	Dimensions
4	2x4s	64" long
9	2x4s	14" long

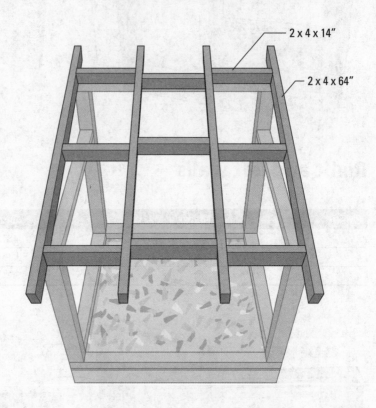

2 x 4 x 14"

2 x 4 x 64"

Nest boxes

Quantity	Material	Dimensions
2	Sheets of plywood	41" x 11"
4	Sheets of OSB	11" x 11"
2	2x4s	13" long
1	2x4	12½" long

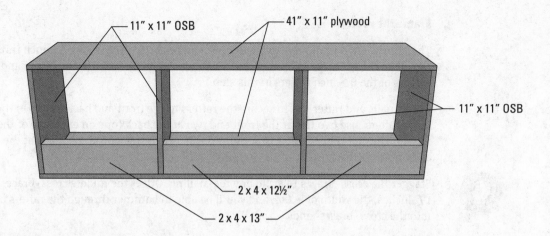

11" x 11" OSB

41" x 11" plywood

11" x 11" OSB

2 x 4 x 12½"

2 x 4 x 13"

Assembling the Coop

Using the preceding cut list tables and illustrations as guides, here are the steps to follow when constructing the Minimal Coop:

1. **Construct the floor.**

    The inner floor joists are centered on the 16-inch and 32-inch marks of the 48-inch pieces. Use 12d nails (2 in each joist end) to assemble the subfloor framework using basic framing techniques, as found in Chapters 5 and 7.

    Add the OSB decking over the top of the completed framework. Fasten it down with 7d nails, one every 6 inches or so. (For more details, see Chapter 7.)

    TIP

    We designed the Minimal to sit atop four concrete pier blocks, most of which have recesses to accept 2x4s laid on end. You should install the finished subfloor on the pier blocks at this stage, so you can make any necessary adjustments before building the rest of the structure. (Be sure the blocks are where you want the completed coop to be placed in your yard.) Then leave the coop on top of the blocks for the duration of the build. Once the coop is complete, it will be too heavy to lift and too difficult to make any tweaks if it doesn't fit properly on the pier blocks. (Flip to Chapter 6 for more information on using concrete pier blocks.)

2. **Frame the walls.**

    Using 12d nails (2 in each stud end) and the methods laid out in Chapter 7, build and secure the walls in the order prescribed by the cut list: front, back, right, left. Be sure that the two tallest 4x4s (the ones that are 45 inches) are "bookends" on the side of the coop that you want to be the front.

    As you fit the right and left walls in place, fasten them to the front and back walls with 2–3 additional 12d nails through the 2x4 and into the 4x4.

3. **Frame the roof.**

   The nine 14-inch pieces that serve as cross-braces between the long 64-inch rafters can all be cut out of scrap lumber that's left over from the two previous steps and the cutting of the 64-inch rafters in this step.

   Use 7d nails and rafter ties to secure the rafters to the front and back walls. On the inner rafters, use two ties at the front and two at the back (one on each side of the rafter). The end rafters will accept rafter ties on the inside of the coop, one at the front and one at the back.

TIP

   Stagger the cross-braces to avoid any toe-nailing. Offset the middle cross-braces by 1½ inches, the width of a 2x4, and you'll be able to hammer through the rafters and into the cross-braces' ends.

4. **Build the nest boxes.**

   Use leftover plywood from the walls to make the top and bottom of the nest box unit and leftover OSB from the floor to create the two dividers. The front ledges are also made from scrap 2x4 pieces. Use them as spacers to help you place the dividers. We used the two 13-inch pieces on the end nests, and the slightly smaller 12½-inch piece in the middle. (Your birds probably won't notice the difference.)

   Use 7d nails to fasten all the pieces together. The thickness of the plywood and OSB should allow you to safely nail through the edges, but for added stability, you could use your last remaining scraps of 2x4 as inner nailing cleats.

WARNING

   You can wait to build the nest boxes last, but they might be harder to get into the coop once the exterior walls are in place.

5. **Attach the exterior walls on the front, back, and one side; create the "door" on the other side.**

   Use 7d nails to fasten the plywood walls to the stud frames on the coop's front and back walls, along with one of the side walls. Go with one nail every 6 inches or so around the perimeter of each wall, making sure you're penetrating the studs behind the plywood. (Chapter 8 has more on installing exterior sheathing.)

WARNING

   Do not attach the second side wall directly to the coop's framework, because it serves as the door and it needs to be mounted with hinges in order to open and close. (In our design, it's the right wall, but you could just as easily make it the left wall of your coop.)

   Hold the side in place so that it closes off the side wall and use the strap hinges to fasten this side to the back wall. Two hinges should do the trick, each one just a few inches away from the door's top and bottom. Be sure that the wood screws that come with the hinges are properly anchored in the 4x4 corner stud, and that they don't come through the relatively thin plywood, where they won't provide as much strength and their sharp points could be a danger to your flock. A simple chain lock (like you'd find on the inside of a hotel room door) attaches easily to the front wall to keep the door closed. Use the wood screws that come with the lock, anchoring them securely into the framing studs. How high or low on the door you place the lock is entirely up to you, but a mounting location near the door's center point may be best for thwarting predators.

TIP

If using a chain lock, stretch the chain far enough on the front wall that it holds the coop door closed tightly when locked. Too much slack might allow a crafty predator to reach in and help himself. Play with a few placement options before making the final attachments, to make sure you'll be able to easily lock and unlock the door as needed.

6. **Secure wire mesh above the side walls.**

Cut the rectangular wire mesh pieces into triangular panels that will fully cover the gable openings. For a piece of fencing that's 12 inches x 48 inches, this requires just one diagonal cut with tin snips, as described in Chapter 3. (Alternatively, you can neatly trim along the mesh squares so that there are no sharp edges, as shown in the figure in this section.)

Use fencing staples (about 20 per piece of fencing) to secure the panels in place above the side walls.

7. **Put down the roofing.**

We used corrugated metal roof panels on the Minimal, and needed two pieces that overlap in the middle to cover the 48-inch width. The 64-inch length allows for plenty of overhang on the front and back. Depending on your panel size, you may be able to put some overhang on either side as well, to help keep things dry inside the coop during a rainstorm.

REMEMBER

Be sure to use whatever fasteners are recommended for your particular roofing panels. See Chapter 8 for more about corrugated roofing.

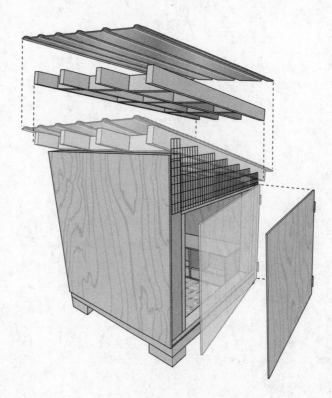

Chapter **13**
The Alpine A-Frame

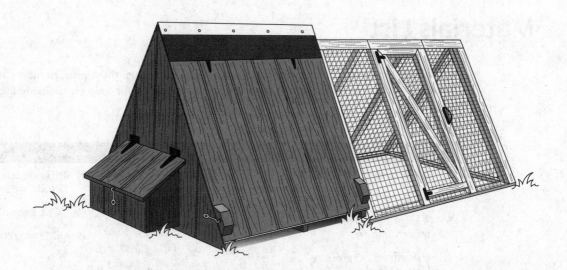

A simple A-frame coop fits many chicken owners to a T. Its compact design takes up a minimum of yard space yet is ideally scaled for a small starter flock. A-frames often require just a modest list of basic building materials and can usually be built very quickly, even by a novice DIYer. It's no surprise, then, that an A-frame is one of the most popular coop styles around, and often the perfect choice for the first-time chicken-keeper or coop-builder. (To see how an A-frame stacks up against other coop styles, check out Chapter 2.)

Vital Stats

>> **Size:** At roughly 4 feet wide and 10 feet long, it occupies just a small footprint in your yard and can easily be tucked into an out-of-the-way corner if needed. At its peak, it stands 46 inches high.

>> **Capacity:** The shelter measures 16 square feet, and the run takes up 24 square feet, making this coop just right for two to four birds.

>> **Access:** One entire side of the A-frame roof acts as a door that can be lifted open for easy cleaning and access, and then latched closed. Chickens have their own door to the run which can also be left open or latched shut.

>> **Nest boxes:** Two nest boxes built into the exterior wall of the shelter share a top-opening (and latchable!) lid for hassle-free egg-gathering.

>> **Run:** A large door positioned in the center of the run allows you to reach in to catch a chicken without having to crawl all the way inside.

>> **Degree of difficulty:** The Alpine is one of the smallest coops featured in this book, and is therefore one of the easiest to construct. Any homeowner with even a basic set of tools and rudimentary skills should be able to build this coop with relatively little trouble.

>> **Cost:** Prices will vary considerably, of course, but assuming that you have all the tools you need, you could easily purchase all of the materials to build this coop from scratch for well under $300.

Materials List

The following is a list of the materials that we used to construct the Alpine. In most cases, substitutions may be made as needed or desired. (See Chapter 4 for help on choosing appropriate alternatives.)

Lumber		Hardware		Fasteners	
3	4′ x 8′ sheets of T1-11	1	4′ x 16′ panel of 1″ x 2″ wire mesh	70	2½″ wood screws or 12 nails
1	4′ x 4′ sheet of ⁷⁄₁₆″ OSB	1	piece of metal 6″ x 49″ (bent lengthwise in the middle)	140	1¼″ screws or 7d nails
3	4′ lengths of 2x4 pressure-treated lumber	1	piece of rubber 8″ x 48½″	120	1¼″ fencing staples
3	6′ lengths of 2x3 pressure-treated lumber	8	5″ T-style hinges		
18	6′ lengths of 2x3 lumber	3	hook-and-eye latches		
1	4′ length of 2x2 lumber	1	gate latch		
2	6′ lengths of 1x3 pressure-treated lumber				
2	6′ lengths of 1x3 lumber				
1	8′ length of 1x2 lumber				

Cut List

These sections break down how the materials should be cut to create the appropriate–sized pieces you need to assemble the Alpine as shown. (Refer to Chapter 5 for a crash course in cutting lumber.)

REMEMBER

This coop features several pieces that are cut on an angle. You'll see boards throughout the cut list that specify L.P. or S.P.

» *L.P.* stands for "long point," where the board's listed length refers to the tip-to-tip measurement at the board's longest point, with the angle making one side shorter.

» *S.P.* is an abbreviation for "short point." When you see it, know that the board's listed length measures just to the end's shortest point, meaning that the long end of the board will be greater than the cut list's specified length.

For example, a 2x4 listed as 36 inches S.P. with a 45-degree angle will actually be almost 39½ inches long at its longest point. (See Chapter 5 for details on how to lay out angles before cutting to ensure accuracy.)

Shelter floor

Quantity	Material	Dimensions	Notes
1	OSB	48" x 47"	
3	Pressure-treated 2x4s	48" long	

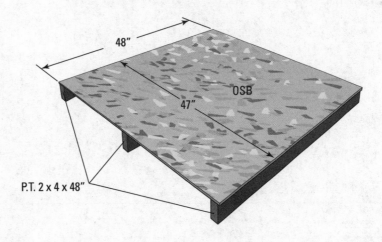

48"

47"

OSB

P.T. 2 x 4 x 48"

Gable 1 (nest box side)

Quantity	Material	Dimensions	Notes
1	Sheet of T1-11	48" wide x 42" high	From top center point, cut both sides at a 30-degree angle to leave vertical edge 1⅝" high at each bottom corner
2	2x3s	47⅛" long L.P. to L.P.	One end trimmed at a 60-degree angle and the other end trimmed at a 33-degree angle
1	2x3	24" long	
2	2x3s	14¼" long	

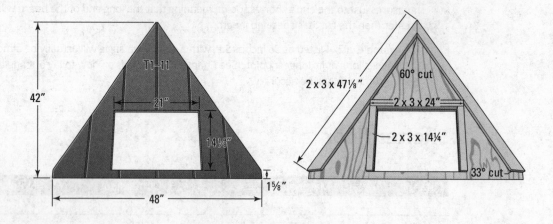

Gable 2 (run side, interior)

Quantity	Material	Dimensions	Notes
1	Sheet of T1-11	48" wide x 42" high	From the top center point, cut both sides at a 30-degree angle to leave a vertical edge 1⅝" high at each bottom corner
2	2x3s	47⅛" long L.P. to L.P.	One end trimmed at a 60-degree angle and the other end trimmed at a 33-degree angle
2	2x3s	16" long	
2	2x3s	14" long	
2	1x2s	13¼" long	
1	1x2	11" long	

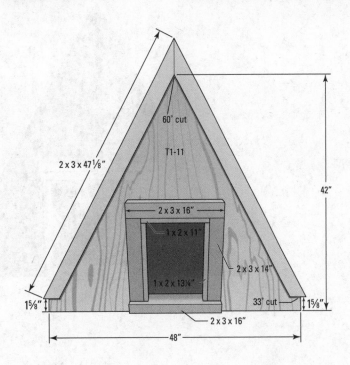

Gable 2 (run side, exterior) with run door

Quantity	Material	Dimensions	Notes
1	Sheet of T1-11	10½" wide x 13½" high	
1	Pressure-treated 1x3	17⅝" long L.P. to L.P.	Ends trimmed at a 15-degree angle
2	Pressure-treated 1x3s	17⅛" long	
1	Pressure-treated 1x3	10⅞" long	
2	Pressure-treated 1x3s	10½" long	
2	Pressure-treated 1x3s	9" long	

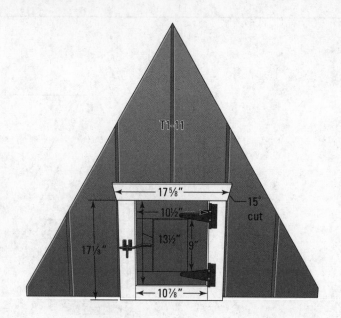

Roof panel 1 (fixed side)

Quantity	Material	Dimensions	Notes
1	Sheet of T1-11	48" wide x 48" high	
1	Sheet of T1-11	48" wide x 6¼" high	
1	2x4	44¾" long	Cut down both long sides at a 30-degree angle. (This creates a 44"-long board that's 3½" on the bottom and 1⅞" on the top.)
2	2x3s	44¾" long	

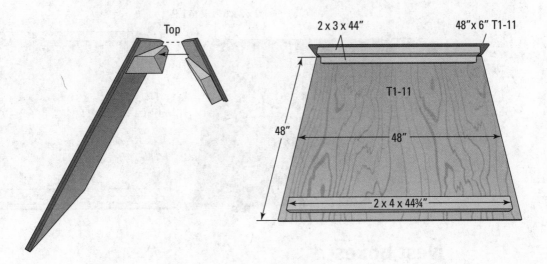

Roof panel 2 (hinged door)

Quantity	Material	Dimensions	Notes
1	Sheet of T1-11	48" wide x 41½" high	
2	2x3s	43¾" long	
2	2x3s	35" long	
2	2x3s	6" long L.P. to L.P.	Both ends trimmed at a 45-degree angle

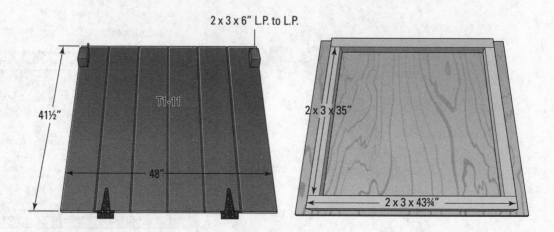

Nest boxes

Quantity	Material	Dimensions	Notes
1	Sheet of T1-11	23¼" wide x 13" high	Cut notches into back corners that measure 1" wide x ¾" high (nest boxes' lid)
1	Sheet of T1-11	20⅞" wide by 7½" high	Top edge trimmed at a 30-degree angle (nest boxes' front)
1	Sheet of T1-11	20" wide x 11½" high	(Nest boxes' bottom)
1	Sheet of T1-11	11¾" wide x 12¾" high (back) and 6" high (front)	(Nest boxes' divider)
2	Sheets of T1-11	11¼" wide x 14" high (back) and 7⅛" high (front)	(Nest boxes' sides)
1	1x3	11¾" long	(Inner blocking)
2	1x3s	10" long	(Inner blocking)
2	1x3s	6½" long	One end trimmed at a 30-degree angle (inner blocking)

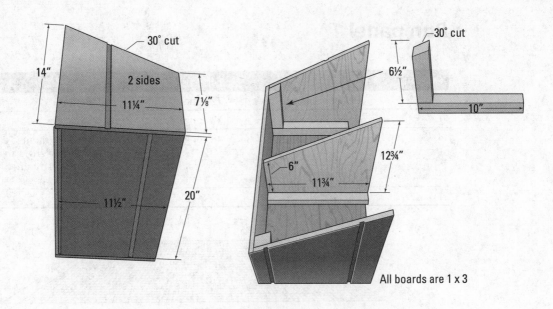

30° cut

14"

2 sides

11¼"

7⅛"

11½"

20"

30° cut

6½"

10"

12¾"

6"

11¾"

All boards are 1 x 3

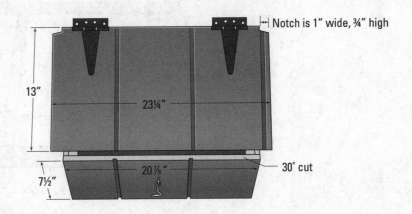

Notch is 1" wide, ¾" high

13"

23¼"

7½"

20⅞"

30° cut

Roost

Quantity	Material	Dimensions	Notes
1	2x2	47" long	Not shown

Run panel 1

Quantity	Material	Dimensions	Notes
1	Pressure-treated 2x3	72" long	(Bottom)
1	2x3	72" long	(Top)
4	2x3s	48" long L.P. to L.P.	One end trimmed at a 60-degree angle and the other end trimmed at a 30-degree angle
1	Wire panel	48" x 71"	

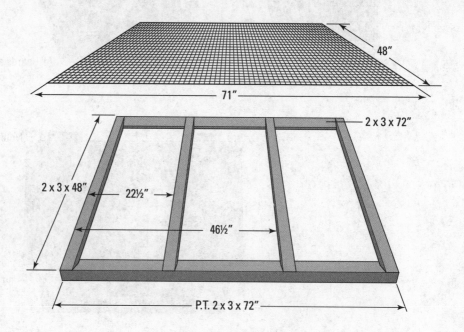

Run panel 2 with door

Quantity	Material	Dimensions	Notes
1	Pressure-treated 2x3	72" long	(Bottom)
1	2x3	72" long	(Top)
4	2x3s	48" long L.P. to L.P.	One end trimmed at a 60-degree angle, and the other end trimmed at a 30-degree angle
1	2x3	42⅞" long L.P. to S.P.	Both ends trimmed at a 22½-degree angle
2	2x3s	40" long	
2	2x3s	21" long	
1	2x3	7" long	(Blocking for latch)
1	Wire panel	24" x 48"	(Shelter-end panel)
1	Wire panel	23" x 48"	(Far-end panel)
1	Wire panel	20" x 40"	(Door panel)

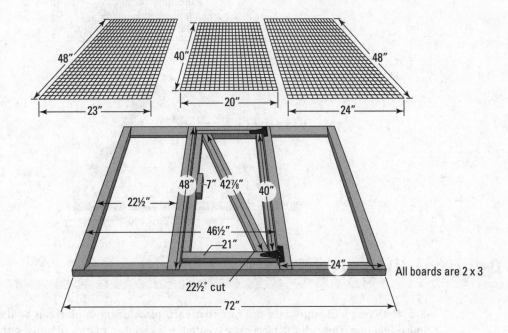

All boards are 2 x 3

Run gable

Quantity	Material	Dimensions	Notes
1	Pressure-treated 2x3	46" long	(Bottom)
2	2x3s	44⅛" long L.P. to S.P.	Start with a 2x3 that's 46½" long. Trim the bottom end to a 30-degree angle. Trim the top end to a 60-degree angle, so that if you rest the 30-degree end flat on the ground, the 60-degree cut is plumb. Now measure up from the bottom of the 2x3 and mark the 44⅛" spot. Make a 30-degree cut at this mark (the cut should run horizontally, or parallel with the bottom cut).
1	Wire panel	47" by 40"	Cut to 40" high triangle

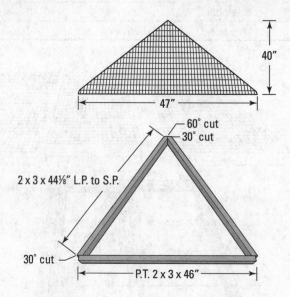

Assembling the Coop

Rookie DIYers will appreciate the fact that each piece of the Alpine can be assembled as an individual unit. Once all the pieces are created, it's simply a matter of putting them together.

1. **Begin with the shelter.**

    Secure the shelter floor to its 2x4 runners with 1¼-inch screws or 7d nails.

    Assemble the triangular gables, using 2½-inch screws for the framing and 1¼-inch screws to attach the T1-11. On the nest box side, cut an opening that is 21 inches wide and 14⅛ inches tall. The opening should be located 1⅝ inches up from the bottom edge of the T1-11 and centered. Attach the gables in place at the floor with 2½-inch screws.

Follow by building roof panel 1 (the fixed side), using 1¼-inch screws to attach the framing members to the T1-11. Then fasten the base to the shelter floor with 2½-inch screws. The 2x3 bracing inside the roof panel and gables should help "lock" the pieces together while you screw everything in place.

Assemble roof panel 2 by securing the framing to the T1-11 with 1¼-inch screws. Attach roof panel 2 (the door) to roof panel 1 with two T-style hinges; place the hinges about 10 inches from the door's edges, and use the screws provided. (If the hinge screws will penetrate the T1-11, consider substituting smaller screws or bolts and nuts.) The base of the roof door shouldn't be attached to anything, so that the door may open freely. Two hook-and-eye latches screwed into small wooden cleats and the shelter floor can lock the door in the closed position.

TIP

Our design features a piece of metal at the shelter peak to keep rainwater from running down into the coop. We had a local machine shop bend a length of steel in order to meet the roof at the proper angle. You should be able to find alternatives at your neighborhood building supply store. We recommend against using fasteners to attach the flashing to the coop, because any hole drilled through the metal flashing creates a possible leak point. Use heavy-duty construction adhesive to glue the flashing to the coop. If the flashing is heavy enough to stay put on its own, you can simply allow it to rest in place — no permanent attachment is necessary.

TIP

A short length of rubber is draped over the hinged door's seam in our design. This serves as a flexible strip of flashing that lets rainwater from the metal peak bypass the door's seam and shed right off onto the roof panel itself while still allowing the door to be opened. A scrap piece of rubber pond liner or a swatch of heavy-duty plastic sheeting should do the trick on your coop. Use short screws or an adhesive appropriate for rubber to affix the top edge of the rubber strip to the coop at the peak. (If you use screws that penetrate the rubber, be sure they're positioned so that the metal flashing piece covers the screw heads to avoid leaks.) Leave the bottom edge loose so that it can move freely with the door when opened.

2. **Add the nest boxes and roost.**

    Use the diagrams that appear in the earlier section "Nest boxes" to cut and assemble the pieces for the nest boxes; use 1¼-inch screws.

    The box itself is then inserted into the appropriate gable hole. Fasten the nest box in place with 1¼-inch screws on the inside of the shelter, drilled through the 2x3 blocking around the gable hole.

    The nest box lid is secured in place to the coop's exterior wall with two T-style hinges, fastened with the screws provided. A hook-and-eye latch, with one end threaded into the lid and the other end threaded into the front of the box, locks it shut.

    With the shelter complete, you can add the roost bar. The length specified in the cut list should allow the bar to span the coop's two gable ends. Position the bar above the chicken door's interior blocking trim, and use small screws or nails to fasten it in place.

3. **Assemble the run.**

Build the run panels and door, along with the run gable, according to the diagrams that appear earlier in this chapter. Then staple wire mesh in place at each section; the staples should be spaced about 8 inches apart.

The panel without the door can be covered by one large piece of fencing, but treat run panel 2 (with the door) as three distinct panels, each needing its own section of fencing.

Fasten the run panels to one another with 2½-inch screws, using the triangular run gable to assist you in holding the pieces together. Secure the run to the shelter with additional 2½-inch screws.

Add a latch to the run's door panel. We used a self-locking model like you would typically see on a fence gate, but almost any style you prefer (barrel bolt, hook-and-eye, hasp) can be made to work. Use the hardware provided to fasten it to the run's door panel.

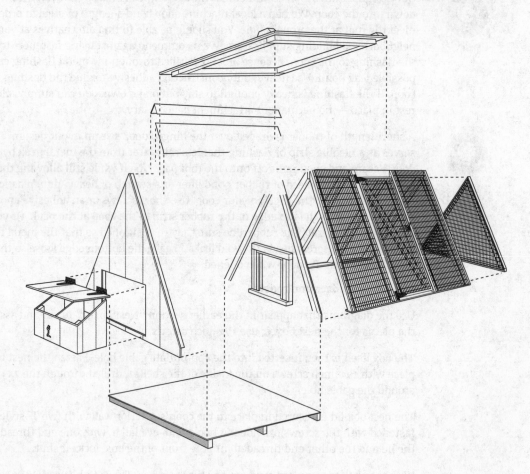

Chapter 14
The Urban Tractor

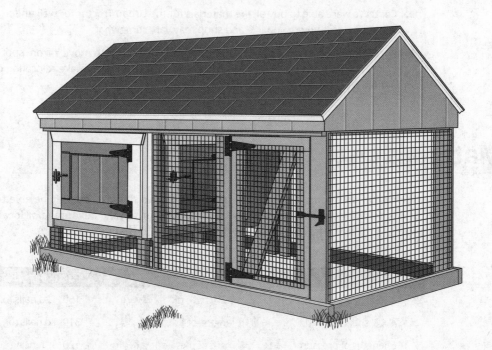

For the city-dweller or suburbanite who's interested in raising a backyard flock, space is often the key concern. Having pet chickens scurry around and getting eggs fresh from the source every morning may sound great, but not if it means giving up your vegetable garden or your kids' play-space. Your solution? Check out this small tractor coop. It takes up a tiny portion of property and is completely self-contained. And when the birds are ready to work a new patch of ground (or your kids are ready to kick off the Neighborhood Cup soccer tourney), you can just drag it to a new spot in the yard.

Vital Stats

» **Size:** The Urban Tractor sits on a footprint that's just 3 x 6 feet, making it perfect for the small, in-town backyard. The peak of the roof is about 5 feet off the ground.

» **Capacity:** A small flock of two to three birds will feel right at home in this coop, with a shelter that provides just over 6 square feet of living space.

» **Access:** The shelter has a small hatch on one exterior wall that allows the caretaker to reach inside. A chicken door can be left open for free access to the run. The enclosed run also features a large door that allows entry.

» **Nest boxes:** An exterior-mounted nest box provides two stalls for the birds and has a latchable lid for easy egg collection.

» **Run:** The Urban Tractor's run measures about 16 square feet, all covered by a roof. A portion of the run reaches underneath the shelter itself, giving your chickens a shady spot to chill out on hot summer days.

» **Degree of difficulty:** This coop is only slightly more advanced than the Alpine A-frame in Chapter 13, because the introduction of vertical walls means some extra framing. This is still a great coop for the novice builder, though.

» **Cost:** We were able to buy all the materials for the Urban Tractor for well under $400, although prices where you live and shop may vary somewhat.

» **Portability:** The whole point of a tractor coop is being able to move it from spot to spot. Ours includes a sturdy lumber base and heavy tow chains that make relocation of the coop a snap. (You could add wheels, too!)

Materials List

Here's what you'll need to build the Urban Tractor as shown. You may be able to make substitutions in order to use materials you already have or to take advantage of local availability or pricing. (Chapter 4 has more on selecting suitable alternative materials.)

Lumber		Hardware		Fasteners	
3	4' x 8' sheets of ½" OSB	1	3' x 12' length of 1" x 2" wire mesh	150	12d nails (or 3" screws)
2	4' x 8' sheets of T1-11	3	10' lengths of drip edge	310	7d nails (or 1¼" screws)
7	6' lengths of pressure-treated 2x4 lumber	1	8' length of metal corner trim	120	1" fencing staples
		28	36" shingles	160	1" galvanized roofing nails
3	6' lengths of 2x4 lumber	2	48" lengths of heavy $\frac{3}{16}$" chain		
15	8' lengths of 2x3 lumber	4	¼" x 1½" lag screws		
12	8' lengths of 1x3 lumber	8	5" T-style hinges		
		4	gate latches		
		1	hook-and-eye latch		

Cut List

As with all our coops, we break the Urban Tractor down into easy-to-build sections. The following sections demonstrate how to cut your lumber and assemble each section, so that you can then put the sections together to finish the coop. (For more on how to cut lumber safely and accurately, refer to Chapter 5.)

REMEMBER

This coop features a few pieces that are cut on an angle. Some boards in the cut list specify L.P. or S.P.:

» *L.P.* stands for "long point," where the board's listed length refers to the tip-to-tip measurement at the board's longest point, with the angle making one side of the board shorter.

» *S.P.* is an abbreviation for "short point." When you see it, know that the board's listed length measures just to the end's shortest point, meaning that the long end of the board will be greater than the cut list's specified length.

For example, a 2x4 listed as 36 inches S.P. with a 45-degree angle cut into one end will actually be almost 39½ inches long at its longest point: 36 inches in actual length, plus the 3½ inches in width of the board. (See Chapter 5 for tips on ensuring accuracy by laying out your angles before cutting.)

This coop features several narrow pieces of lumber that must be *ripped*, or carefully cut lengthwise from wider lengths of stock lumber. Refer to Chapter 5 for more on how to safely and accurately perform a rip cut.

Tractor base and shelter floor

Quantity	Material	Dimensions	Notes
1	Sheet of OSB	36" x 25¼"	
2	Pressure-treated 2x4s	70" long	
2	Pressure-treated 2x4s	33" long	
4	Pressure-treated 2x4s	15" long	
2	2x3s	33" long	
2	2x3s	25¼" long	

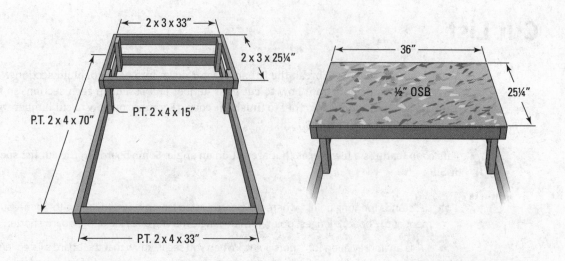

Left wall

Quantity	Material	Dimensions	Notes
1	Sheet of T1-11	25" wide x 23" high	
2	2x3s	25¼" long	
2	2x3s	17" long	

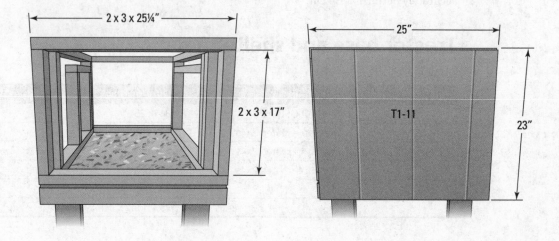

Right wall

Quantity	Material	Dimensions	Notes
1	Sheet of T1-11	25" wide x 23" high	
2	2x3s	25¼" long	
1	2x3	19¼" long	
4	2x3s	17" long	
1	1x3, ripped to 2" wide	26⅛" long S.P. to S.P.	Both ends cut at a 15-degree angle (doorway trim)
1	1x3	25⅞" long	(Doorway trim)
2	1x3s	15½" long	(Doorway trim)

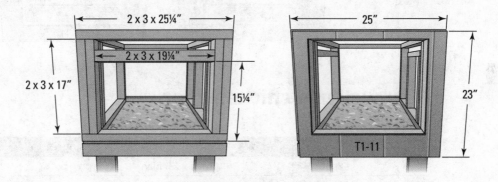

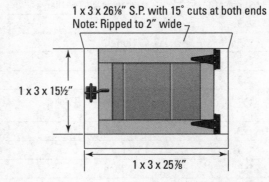

1 x 3 x 26⅛" S.P. with 15° cuts at both ends
Note: Ripped to 2" wide

Front wall

Quantity	Material	Dimensions	Notes
1	Sheet of T1-11	36¼" wide x 23" high	
2	2x3s	31" long	
1	2x3	24⅜" long	
4	2x3s	17" long	

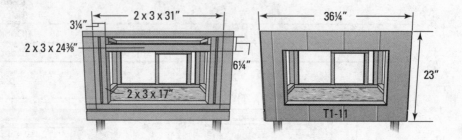

Rear wall and chicken door

Quantity	Material	Dimensions	Notes
1	Sheet of T1-11	36¼" wide x 23" high	
1	Sheet of T1-11	10½" wide x 13⅝" high	
2	2x3s	31" long	
2	2x3s	17" long	
2	2x3s	14¼" long	
1	2x3	14" long	
1	1x3	16⅝" long S.P. to S.P.	Both ends cut at a 15-degree angle (doorway trim)
1	1x3	16¼" long	(Doorway trim)
2	1x3s	15" long	(Doorway trim)
2	1x3s	13¼" long	(Slam strips)
1	1x3	10⅞" long	(Slam strip)
2	1x3s	10½" long	
2	1x3s	9" long	

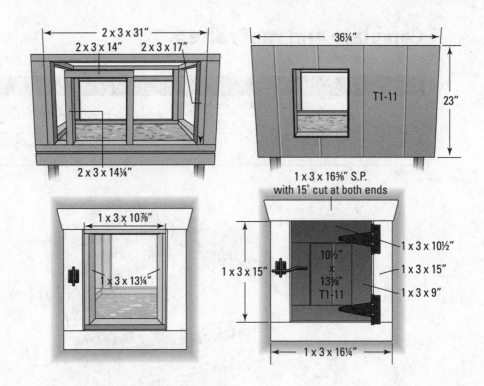

2 × 3 × 31″

2 × 3 × 14″   2 × 3 × 17″

2 × 3 × 14¼″

36¼″

T1-11

23″

1 × 3 × 16⅝″ S.P.
with 15° cut at both ends

1 × 3 × 10⅞″

1 × 3 × 13¼″

1 × 3 × 15″

10½″
×
13⅝″
T1-11

1 × 3 × 10½″

1 × 3 × 15″

1 × 3 × 9″

1 × 3 × 16¼″

Run posts and framing

Quantity	Material	Dimensions	Notes
3	Pressure-treated 2x4s	35½″ long	(Posts)
2	2x3s	42¾″ long	
2	2x3s	35½″ long	(Posts)
1	2x3	30½″ long	

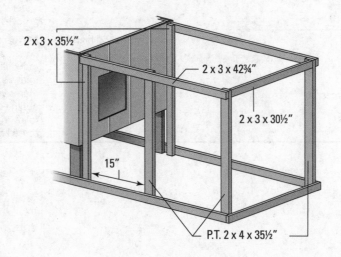

2 × 3 × 35½″

2 × 3 × 42¾″

2 × 3 × 30½″

15″

P.T. 2 × 4 × 35½″

Cap plate and roof rafters

Quantity	Material	Dimensions	Notes
2	2x4s	69" long	
2	2x4s	30" long	
10	2x3s	26½" long L.P. to L.P.	Both ends trimmed at a 45-degree angle

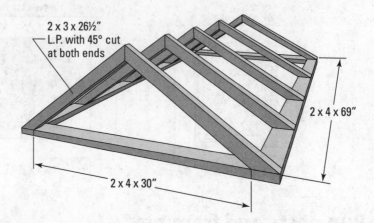

2 x 3 x 26½"
L.P. with 45° cut
at both ends

2 x 4 x 69"

2 x 4 x 30"

Roof, fascia, and gables

Quantity	Material	Dimensions	Notes
2	Sheets of OSB	26½" x 69"	
2	Sheets of T1-11	69" long x 3" high	Although the width of a full sheet of T1-11 is 48", there are no visible grooves in this piece. You can turn a sheet sideways and cut between the grooves to get the given dimensions.
2	Sheets of T1-11	38" wide x 22" high	Make a mark on the left and right edges, 3" up from the bottom. Make a mark along the top edge at the piece's midway point, or 19". Connect the marks with pencil lines to create an outline of the shape shown in the following figure, and cut along these lines.

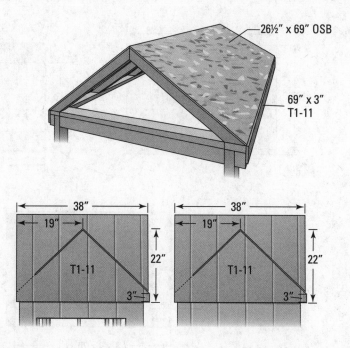

26½" x 69" OSB

69" x 3"
T1-11

38"

19"

22"

T1-11

3"

38"

19"

22"

T1-11

3"

Access door

Quantity	Material	Dimensions	Notes
1	Sheet of T1-11	18⅝" wide x 13¼" high	
2	2x3s	18¼" long	(Interior frame)
2	2x3s	7¾" long	(Interior frame)
2	1x3s	20¼" long	
2	1x3s	9½" long	

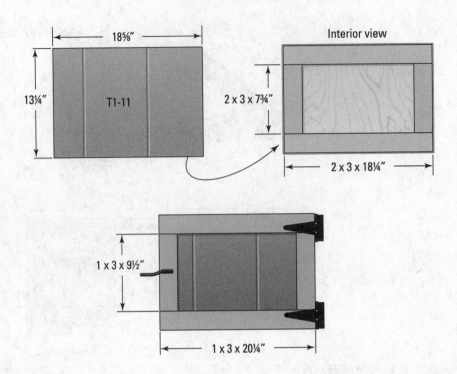

Nest boxes

Quantity	Material	Dimensions	Notes
1	Sheet of T1-11	26" wide x 15" high	1" x 1" notches cut into top corners and top edge trimmed at a 30-degree angle (lid)
2	Sheets of T1-11	11¾" wide x 13¾" high on one side and 9" high on opposite side	(Ends)
1	Sheet of T1-11	11¾" wide x 12" high on one side and 7½" high on opposite side	(Middle divider)
1	Sheet of T1-11	10½" wide x 24" high	(Bottom)
2	1x3s	23¼" long	(Lid frame)
4	1x3s	11¾" long	(Interior framing)
4	1x3s	6¾" long	(Interior framing)
4	1x3s	6½" long	(Interior framing)
2	1x3s	5½" long	(Lid frame)

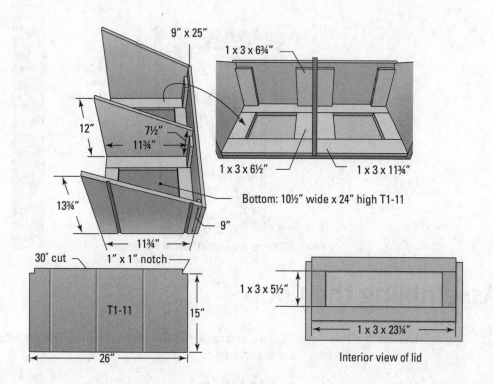

9" x 25"

1 x 3 x 6¾"

12"

7½"

11¾"

1 x 3 x 6½"

1 x 3 x 11¾"

13¾"

9"

Bottom: 10½" wide x 24" high T1-11

11¾"

30° cut

1" x 1" notch

1 x 3 x 5½"

T1-11

15"

1 x 3 x 23¼"

26"

Interior view of lid

Run door and chicken ramp

Quantity	Material	Dimensions	Notes
2	2x3s	29" long	
1	2x3	25⅞" long L.P. to S.P.	Both ends cut at a 21-degree angle
2	2x3s	12½" long	
1	2x3	7" long	
2	1x3s	36" long S.P.	One edge cut to a 30-degree angle
4	1x3s	5" long	

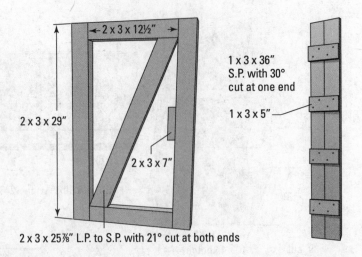

2 x 3 x 12½"

1 x 3 x 36"
S.P. with 30°
cut at one end

2 x 3 x 29"

1 x 3 x 5"

2 x 3 x 7"

2 x 3 x 25⅞" L.P. to S.P. with 21° cut at both ends

Assembling the Coop

Here are the steps necessary to assemble the Urban Tractor. Use the preceding cut list and figures as guides in putting the various pieces together.

1. **Construct the tractor base and coop floor.**

    Be sure that the 33-inch pieces (2x4s on the tractor base; 2x3s on the shelter floor frame) are fastened in between the longer 70-inch 2x4s on the tractor base and the 25¼-inch 2x3s on the shelter floor. Use a pair of evenly-spaced 12d nails at each joint.

    Build the tractor base first, then the shelter floor as a separate piece. Nail the OSB flooring down to the shelter floor frame with 7d nails every 6 inches.

    Fasten the 15-inch 2x4 legs to the underside of the shelter floor frame — one in each corner — with a pair of 12d nails holding each leg fast. Then set the shelter floor on top of the tractor base at one end, with the legs inside the base. Use a pair of evenly-spaced 12d nails to secure each leg to the tractor base.

2. **Build the left wall, then the right wall.**

    The left wall is a basic stud wall; use a pair of evenly-spaced 12d nails to secure each stud end to the top and bottom plates. Fasten the T1-11 paneling to the wall with 7d nails every 6 inches.

TIP

The right wall features doubled-up studs on either end and a horizontal header for the access door. Use the outer 2x3 studs as spacers to set the inside studs on the top and bottom plates. Fasten them in place first with a pair of 12d nails in each end. Then secure the 19¼-inch header between the inner studs, while you can still drive nails through these 2x3s into the header ends; use two evenly-spaced 12d nails. Then set and fasten the outer studs last with more 12d nails, two at each end (into the top and bottom plates) and three through the side (into the inner stud).

On the right wall, position the 19¼-inch access door header so that it sits 15¼ inches up from the finished floor level. This results in an access door opening that measures 19¼ inches wide x 13¾ inches high.

Secure the T1-11 to the outer frame with a 7d nail every 6 inches. Cut the previously mentioned access door hole in the T1-11 siding.

Do not attach the right wall's trim pieces around the doorway opening until after the metal corner trim has been installed in Step 8!

3. **Build the front wall and the rear wall with the chicken door.**

For the front wall, set the two innermost 17-inch 2x3 studs first, at 3¼ inches in from the plate ends as shown in the figures in the earlier section "Front wall"; use two 12d nails driven through the top and bottom plates into the studs' ends. Then secure the 24⅜-inch nest box header so that it sits 6¼ inches down from the top plate as shown (use two 12d nails driven through each of the inner studs into the header's ends). Finish with two outer 17-inch studs on the plate ends, again, using 12d nails.

Fasten the T1-11 paneling to the outer wall, using 7d nails spaced 6 inches apart around the perimeter of both the wall and the large nest box opening. When everything is spaced properly, the hole in the T1-11 for the nest boxes will measure 24½ inches wide by 13¾ inches high. Cut this hole out using techniques specified in Chapter 8.

After constructing the rear wall's outer frame, position the left door frame stud 4½ inches in from the plate ends and secure it with the same 12d nails, driving two through the bottom plate into the stud end. The 14-inch door header will correctly position the right door frame stud; secure them both with more 12d nails, a pair at each joint. This doorway opening should measure 11 inches wide by 14¼ inches high.

Fasten the T1-11 to the stud framing with 7d nails every 6 inches, cut the doorway hole in the paneling, and attach the trim pieces around the doorway opening with7d nails spaced 4 inches apart. Secure the slam strips to the inside of the doorway frame with a 7d nail every 4 inches.

Assemble and install the chicken door as shown in the figures in the earlier section "Rear wall and chicken door." Start with the door panel, and secure the door's trim pieces using 7d nails every 3 inches or so. Use T-hinges to fasten the door to the coop wall, using the hardware provided with the hinges. Finally, attach a door latch to the opposite side of the door using the included hardware so that the door latches easily when closed.

4. **Install the run posts and framing.**

Attach the 2x3 posts that are 35½ inches long at the coop's rear wall. Using two 12d nails at each end, secure them to the tractor base.

Position a 2x4 post in each remaining corner of the tractor base and secure each one with a pair of 12d nails. Secure the final 2x4 post so that there's a 15-inch gap between it and the 2x3 corner post along the coop's right wall.

Complete the run framing with a 42¾-inch 2x3 along the tops of the posts on the right and left walls, and a 30½-inch 2x3 between the tops of the posts on the run's narrow end; use 12d nails (two in each stud) to make the attachments.

5. **Attach the cap plate and construct the roof rafters.**

   Secure the 2x4 cap plate around the top of the structure's perimeter with 12d nails every 6 inches.

   Fasten the rafter 2x3s to the cap plate and each other at the peak. Toe-nail once through the top front and once through the side rear of each rafter end into the cap plate using 12d nails. Use two 12d nails at each rafter peak, driving one nail from each side in a criss-cross configuration.

   Position the roof rafters so they're spaced 17 inches apart on-center.

REMEMBER

6. **Install the roof sheathing, fascia boards, and gable ends.**

   Secure the sheathing to the rafters with a 7d nail every 6 inches around the perimeter and every 12 inches inside both pieces of OSB.

   The fascia boards are cut out of sheets of T1-11, with 3-inch-wide strips being cut from between the factory grooves. (The fascia boards have no grooves and appear as solid pieces of paneling.) Position the fascia boards to sit just under the roof edge and nail them to the cap plate using 7d nails spaced 6 inches apart.

   Cut the triangular gable ends out of larger pieces of T1-11, following the directions in the cut list and the figure in the earlier section "Roof, fascia, and gables." Attach the gable ends to the roof, nailing into the end rafters and cap plate with a 7d nail every 6 inches.

7. **Install drip edge and roof shingles.**

   Use drip edge along all four edges of the roof. Wait until the roof is fully sheathed before cutting your drip edge pieces, so you can get exact measurements. Use one roofing nail at each rafter location (use about four spaced evenly down the side of each gable end) to fasten the drip edge to the roof.

   Install roof shingles, beginning at the bottom edge and working your way up to the peak. Keep the shingles in a straight line as you add rows, and use four roofing nails per three-tab shingle. Make sure that the nailheads will be hidden by the next row of shingles. (Check out Chapter 8 for a more detailed look at shingling a roof.)

8. **Install the metal corner trim on the shelter's corners.**

   Depending on the thickness or gauge of your corner trim material, use heavy-duty adhesive and/or 7d nails to fasten the trim to each corner of the coop.

   You can now use 7d nails to fasten the doorway trim pieces around the access door on the right wall (see Step 2). These pieces will overlap the metal corner trim.

9. **Build and install the access door.**

   Attach interior and exterior trim pieces to the T1-11 using 7d nails every 3 inches or so. Install the door on the coop wall with T-style hinges and a latch, using the screws provided with the hardware.

10. **Assemble and install the nest boxes.**

    Following the layout shown in the figure in the previous "Nest boxes" section, use 1x3 boards and T1-11 pieces to build the nest boxes. Use the 1x3 boards as interior blocking and nailing strips to help hold the T1-11 pieces together.

We strongly recommend using 1¼-inch screws for the nest boxes' construction and installation. The thickness of a 1x3 plus a piece of T1-11 is about 1⅜ inches. A 7d nail will completely penetrate both pieces, either leaving a sharp point sticking out of the next box (where it could poke you) or into the nest box (where it could poke a hen), or requiring you to snip the pointed ends off by hand. The point of a 1¼-inch wood screw will remain safely embedded in the receiving piece of lumber.

Attach the 1x3 boards to the T1-11 pieces using 1¼-inch screws, one at each end of each 1x3 piece. When fastening pieces of T1-11 together, the 1x3s provide a little extra surface area to drive screws into.

Position the nest boxes in the opening in the coop's front wall. Secure the nest boxes in place with the same fasteners, driving them through the nest box walls into the front wall studs. Use two or three screws per side wall and six to eight along the bottom of the nest boxes.

Attach the nest boxes' lid to the coop wall using a pair of T-hinges and the screws provided. The lid should open freely when lifted and rest on the nest boxes when lowered. A hook-and-eye latch will lock the lid; screw the hook into the edge of the lid and the eye into the nest boxes' front wall so the hook can be inserted as needed.

11. **Construct and install the run door.**

   Using 12d nails, toe-nail the pieces together that comprise the door, following the figure in the "Run door and chicken ramp" section. Two to three nails per joint should provide sufficient holding power. (See Chapter 5 for toe-nailing advice.)

   If you doubt your toe-nailing skills, feel free to use mending plates, angle plates, and/or tee plates at the joints to build the run door. (We mention these handy items in Chapter 5.)

   Don't attach the door to the run until after you've fastened the wire in Step 12. This will allow you to position the hinges and latch more precisely.

12. **Cut and install wire mesh to the run.**

   Starting underneath the main shelter's left wall (the one with no openings in it), attach a piece of wire mesh that measures 24 inches wide x 9 inches high. The run wall on the same side gets the largest piece of wire mesh, 41 inches wide x 29 inches high.

   Next, use a piece 32 inches wide x 30 inches high to cover the end of the run opposite the main shelter.

   On the front of the coop, (the end with the nest boxes), fasten a section of wire mesh that measures 32 inches wide x 9 inches high under the shelter wall.

   You'll need three separate pieces of wire mesh for the run's right wall, the side that features the caretaker's access doors to both the shelter and the run. Underneath the shelter wall, use a piece that's been cut to 24 inches wide x 9 inches high. Between the shelter and the run door, use a section that's 19 inches wide x 29 inches high. Finally, use a piece 16 inches wide x 28 inches high for the run door.

   To secure all these wire pieces to the run frame, use fencing staples, one every 2 to 4 inches or so to ensure a tight connection.

13. Fasten the run door to the frame.

Use T-hinges and the screws provided to secure one side to the frame. Position your latch to the short 7-inch block on the opposite side and use the hardware included to secure it to the door and the frame so that it latches when closed.

14. Secure the tow chains to the tractor base.

Each 48-inch length of chain is attached with lag screws to the sides of the tractor base at both the front and rear of the coop.

15. Assemble the chicken ramp.

The long 1x3 boards rest side-by-side and are held together by 1x3 rungs fastened on top. The ramp can be propped in place against the coop floor when the chicken door is open and removed when the door is to be closed. Alternatively, you can use two or three 7d nails (through the coop wall into the ramp's edge) to carefully fasten the ramp in place just below the door; be sure that the door will swing freely and clear the ramp.

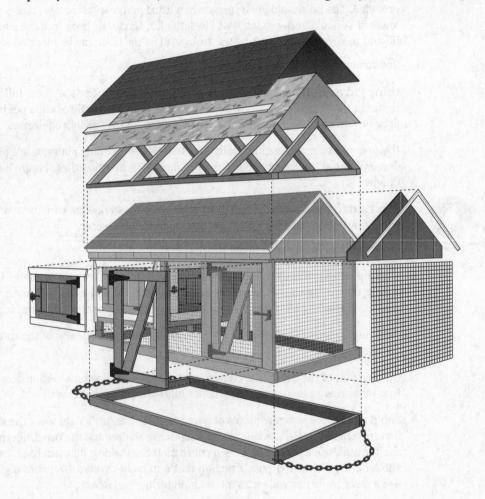

Chapter **15**

The All-in-One

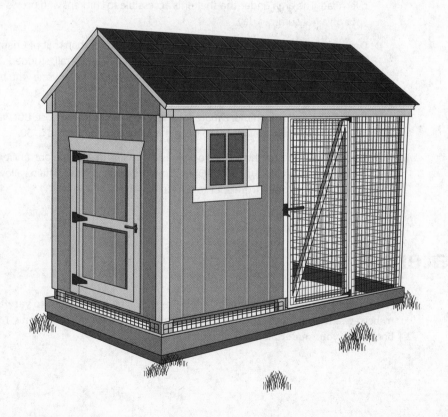

Few things appeal to us as a society like the "all-in-one" concept: a pocketknife that's also a screwdriver, a nail file, a can opener, and a toothpick; the mega-mart superstore that sells Cheez Whiz, blue jeans, and snow tires; a cellphone that doubles as a camera, a GPS device, and an MP3 player. We're all in. For many in-town chicken-owners, the All-in-One chicken coop is just as much of a godsend, thanks to a sizable shelter and a protected run all under the same compact roof. (See Chapter 2 for more on various coop styles.)

Vital Stats

>> **Size:** The All-in-One eats up a patch of ground that's just 8 feet x 4 feet. Its highest point stands less than 7 feet off the ground.

>> **Capacity:** With a shelter that's just a hair under 12 square feet and a run that's just over 26, the All-in-One accommodates four to six birds comfortably.

>> **Access:** The caretaker can fully enter the shelter or the run via doors that are over 3½-feet tall. Chickens have a dedicated door that allows them to come and go between the shelter and the run if left unlatched.

>> **Nest boxes:** Three nest boxes are located inside the shelter. An exterior hatch allows eggs to be collected without physically entering the coop itself.

>> **Run:** The All-in-One's run is completely wrapped by wire mesh and protected by a roof overhead. The area under the shelter is accessible to chickens and provides a cool, shady spot on a hot summer day.

>> **Degree of difficulty:** While the footprint is even smaller than that of Chapter 13's A-frame, the All-in-One ups the ante slightly for the builder by adding tall, studded walls in both the shelter and the run. Even so, this coop should be doable for anyone who has mastered the skills in Chapters 5–10.

>> **Cost:** Taller walls mean more material and, therefore, a higher cost. But we were able to buy everything on the materials list to build the All-in-One for around $750.

>> **Bonus:** This coop combines the roominess of a walk-in with the portability of a tractor coop. (Remember, it *is* called the All-in-One.) Timber skids and heavy chains allow the entire coop to be pulled or towed from spot to spot in your yard.

Materials List

Here's what we used to construct the All-in-One. You may choose to vary the materials you use to better suit your needs, budget, or capabilities. (Check out Chapter 4 for help in going off-book with your materials.)

Lumber		Hardware		Fasteners	
2	4' x 8' sheets of ⅝" OSB	1	12' roll of 48" wire mesh	250	12d nails (or 3" wood screws)
1	4' x 4' sheet of ⅝" OSB	3	10' lengths of drip edge	300	7d nails (or 1¼" wood screws)
3	4' x 8' sheets of T1-11			100	1¼" wood screws
1	4' x 4' sheet of ⅝" plywood	2	8' lengths of metal corner trim	280	1" galvanized roofing nails
2	8' lengths of 4x4 pressure-treated lumber	2	12" x 12" windows	200	1" fencing staples
5	8' lengths of 2x4 pressure-treated lumber	40	36" shingles		
10	8' lengths of 2x4 lumber	2	68" lengths of heavy ³⁄₁₆" chain		
26	8' lengths of 2x3 lumber	4	⁵⁄₁₆" x 1½" lag screws		
1	6' length of 2x2 lumber	9	5" T-style hinges		
1	3' length of 1x6 lumber	4	gate latches		
8	8' lengths of 1x3 lumber	1	rubber, wall-mounted doorstop		
1	2' length of 1x2 lumber				

Cut List

The following sections break the All-in-One coop into smaller pieces. We give detailed instructions on how to cut your lumber so that you can assemble each piece one at a time and then put the pieces together. (Chapter 5 goes over the proper and safe way to make accurate lumber cuts.)

REMEMBER

This coop features several pieces that are cut on an angle. Some boards in the cut list specify L.P. or S.P.:

>> *L.P.* stands for "long point," where the board's listed length refers to the tip-to-tip measurement at the board's longest point, with the angle making one side shorter.

>> *S.P.* is an abbreviation for "short point." When you see it, know that the board's listed length measures just to the end's shortest point, meaning that the long end of the board will be greater than the cut list's specified length.

For example, a 2x4 listed as 36 inches S.P. with a 45-degree angle will actually be almost 39½ inches long at its longest point. (See Chapter 5 for details on how to lay out angles before cutting to ensure accuracy.)

Skids and floor skirting

Quantity	Material	Dimensions	Notes
2	Pressure-treated 4x4s	94" long L.P. to L.P.	Both ends cut at a 50-degree angle
2	Pressure-treated 2x4s	88" long	
2	Pressure-treated 2x4s	44½" long	
4	Pressure-treated 2x4s	12" long	

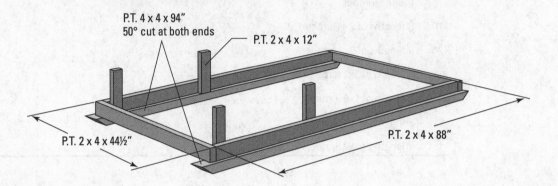

P.T. 4 x 4 x 94"
50° cut at both ends

P.T. 2 x 4 x 12"

P.T. 2 x 4 x 44½"

P.T. 2 x 4 x 88"

Shelter floor

Quantity	Material	Dimensions	Notes
1	Sheet of OSB	39" x 44½"	
2	Pressure-treated 2x4s	44½" long	
1	Pressure-treated 2x4	41½" long	
2	Pressure-treated 2x4s	36" long	

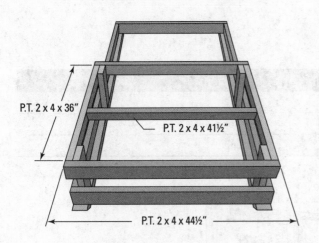

P.T. 2 x 4 x 36"

P.T. 2 x 4 x 41½"

P.T. 2 x 4 x 44½"

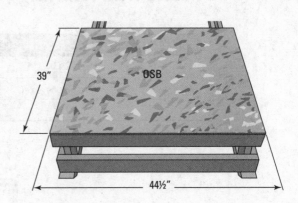

39"

OSB

44½"

Right wall framing

Quantity	Material	Dimensions	Notes
4	2x3s	41" long	
2	2x3s	39" long	
2	2x3s	12¼" long	

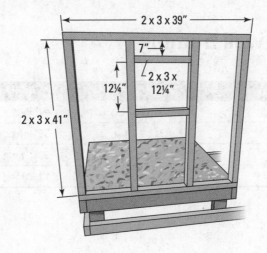

2 x 3 x 39"

7"

2 x 3 x 12¼"

12¼"

2 x 3 x 41"

Front wall framing

Quantity	Material	Dimensions	Notes
4	2x3s	41" long	
2	2x3s	39½" long	
1	2x3	24" long	

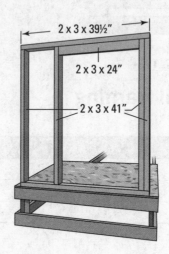

Left wall framing

Quantity	Material	Dimensions	Notes
4	2x3s	41" long	
2	2x3s	39" long	
3	2x3s	32¾" long	
2	2x3s	12¼" long	

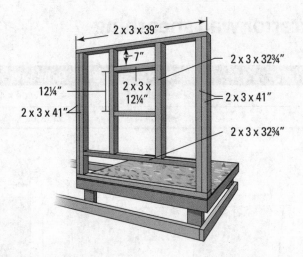

Back wall framing

Quantity	Material	Dimensions	Notes
5	2x3s	41" long	
2	2x3s	39½" long	
1	2x3	11" long	

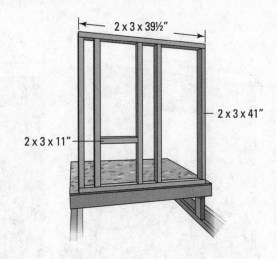

Exterior wall sheathing

Quantity	Material	Dimensions	Notes
2	Sheets of T1-11	45⅜" wide x 48" high	(Front and back walls)
2	Sheets of T1-11	39" wide x 48" high	(Right and left walls)

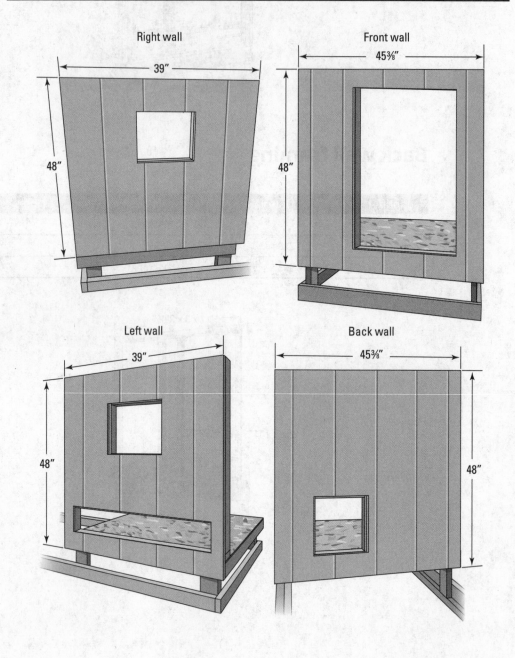

Right wall

39"

48"

Front wall

45⅜"

48"

Left wall

39"

48"

Back wall

45⅜"

48"

Run framing

Quantity	Material	Dimensions	Notes
3	2x4s	56½" long	
2	2x4	49⅞" long	
1	2x4	38⅜" long	
1	2x3	56½" long	
2	2x3s	44¾" long	
2	2x3s	38¼" long	
2	2x3s	21⅜" long	
1	2x2	56½" long	

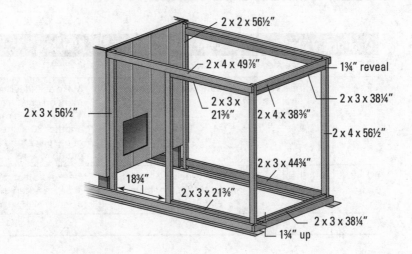

2 x 2 x 56½"

2 x 4 x 49⅞"

1¾" reveal

2 x 3 x 21⅜"

2 x 3 x 38¼"

2 x 4 x 38⅜"

2 x 4 x 56½"

2 x 3 x 56½"

2 x 3 x 44¾"

18¾"

2 x 3 x 21⅜"

2 x 3 x 38¼"

1¾" up

Roof framing and rafters

Quantity	Material	Dimensions	Notes
2	2x4s	89¾" long	
2	2x4s	41⅜" long	
12	2x3s	28" long L.P. to L.P.	One end cut at a 30-degree angle and one end cut at a 60-degree angle

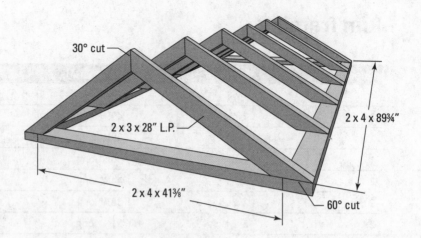

30° cut

2 x 3 x 28" L.P.

2 x 4 x 89¾"

2 x 4 x 41⅜"

60° cut

Roof, gables, and fascia

Quantity	Material	Dimensions	Notes
2	Sheets of T1-11	90" wide x 2½" high	
2	Sheets of T1-11	39⅞" wide x 17" high	Make a 30-degree cut from the 8⅜" mark on one factory-beveled edge to the 17"-high side. From this peak point, make another 30-degree cut to the other end of the piece, and then trim the short end plumb to leave a 3-inch-high edge.
2	Sheets of T1-11	9⅝" wide x 8⅜" high L.P. (at the factory-beveled edge)	One end cut at a 30-degree angle, the short end then trimmed plumb to leave a 3-inch-high edge
2	Sheets of OSB	28" x 90⅝"	

WARNING

The 9⅝-x-8⅜-inch T1-11 pieces must be cut so that the 8⅜-inch sides of each gable will fit together at the factory-beveled edges! This creates the appearance of one vertical "groove" when the pieces are fastened in place on the coop gable.

TIP

These gables will feature vertical grooves, just like the walls that will sit immediately underneath them. If it's important to you that the grooves on the gable line up precisely with the grooves on the wall just below it, you'll have to be very strategic about how and where you make your cuts now out of the full sheets of T1-11.

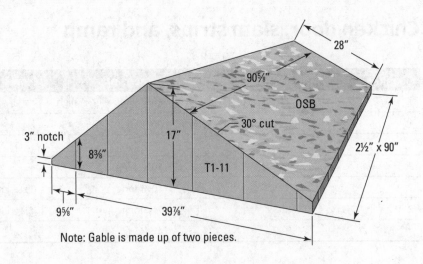

3" notch

8⅜"

17"

90⅝"

28"

OSB

30° cut

2½" x 90"

T1-11

9⅝"

39⅞"

Note: Gable is made up of two pieces.

Shelter access door and slam strips

Quantity	Material	Dimensions	Notes
1	Sheet of T1-11	23½" wide x 38" high	
2	2x3s	30¾" long	(Interior frame)
2	2x3s	21½" long	(Interior frame)
1	2x3	16½" long	(Interior frame)
1	1x3	38⅝" long	(Slam strip; not shown)
2	1x3s	34" long	(Exterior trim)
1	1x3	23⅞" long	(Slam strip; not shown)
2	1x3s	23½" long	(Exterior trim)
1	1x3	18⅜" long	(Exterior trim)

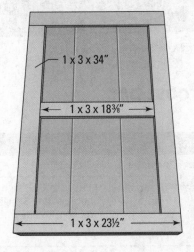

1 x 3 x 34"

1 x 3 x 18⅜"

1 x 3 x 23½"

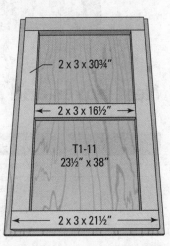

2 x 3 x 30¾"

2 x 3 x 16½"

T1-11
23½" x 38"

2 x 3 x 21½"

Chicken door, slam strips, and ramp

Quantity	Material	Dimensions	Notes
1	Sheet of T1-11	10¾" wide x 11⅞" high	
1	1x6, ripped to 4¾" wide	36" long	
2	1x3s	11½" long	
2	1x3s	10⅞" long	
2	1x3s	10¾" long	(Exterior trim)
2	1x3s	7⅜" long	(Exterior trim)
4	1x2s	4½" long	

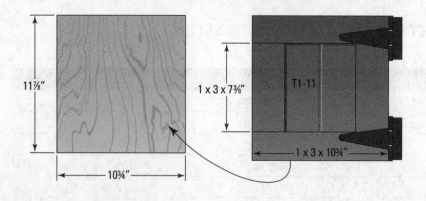

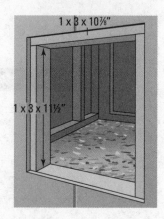

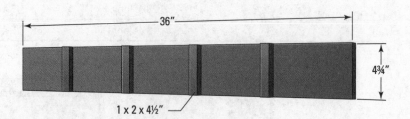

Nest boxes and roost bar

Quantity	Material	Dimensions	Notes
1	Sheet of ⅝" plywood	12¼" x 34"	
4	Sheets of ⅝" plywood	11¾" wide x 12" high on one side and 15" high on opposite side	
1	Sheet of ⅝" plywood	3" x 34"	
1	2x3	44¼" long	(Roost bar; not shown)

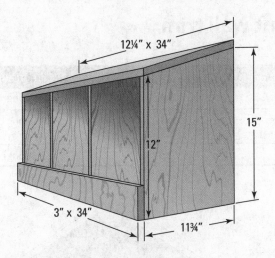

Nest box door and slam strips

Quantity	Material	Dimensions	Notes
1	Sheet of T1-11	32½" wide x 6¼" high	
1	2x4	30½" long	
2	1x3s	33" long	
2	1x3s	5¼" long	

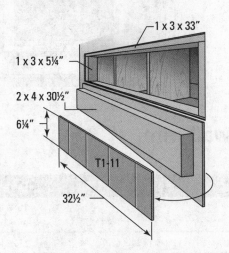

Front wall trim

Quantity	Material	Dimensions	Notes
2	1x3s	42½" long	
1	1x3	29¼" long L.P. to L.P.	Both ends trimmed at a 15-degree angle
1	1x3	23¾" long	

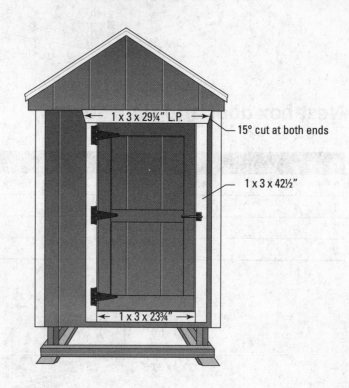

Back wall trim

Quantity	Material	Dimensions	Notes
1	1x3	16¼" long L.P. to L.P.	Both ends trimmed at a 15-degree angle
2	1x3s	15½" long	
1	1x3	11" long	

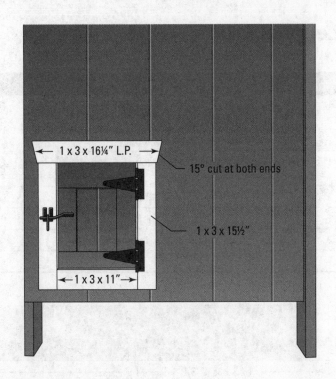

1 x 3 x 16¼" L.P.

15° cut at both ends

1 x 3 x 15½"

1 x 3 x 11"

Right wall trim

Quantity	Material	Dimensions	Notes
1	1x3	19" long S.P. to S.P.	Both ends cut at a 15-degree angle
2	1x3s	16¼" long	
1	1x3	13¾" long	

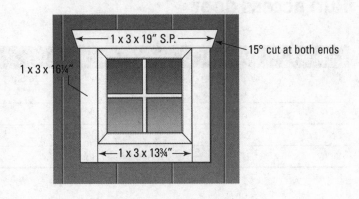

1 x 3 x 19" S.P.

15° cut at both ends

1 x 3 x 16¼"

1 x 3 x 13¾"

Left wall trim

Quantity	Material	Dimensions	Notes
1	1x3	38⅜" long S.P. to S.P.	Both ends cut at a 15-degree angle
1	1x3	33" long	
1	1x3	19" long S.P. to S.P.	Both ends cut at a 15-degree angle (window trim; not shown)
2	1x3s	16¼" long	(Window trim; not shown)
1	1x3	13¾" long	(Window trim; not shown)
2	1x3s	9¼" long	
1	1x3	5" long	

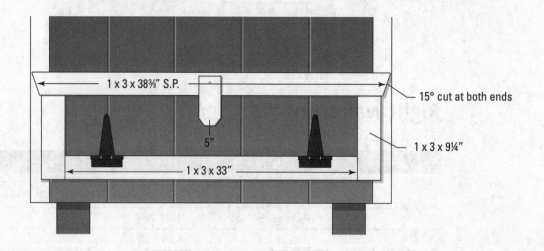

Run access door

Quantity	Material	Dimensions	Notes
1	2x3	44⅞" long L.P. to S.P.	Both ends cut at a 15-degree angle
2	2x3s	43⅝" long	
2	2x3s	18½" long	
1	2x3, ripped to 1½" wide	7" long	

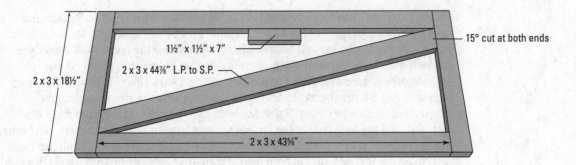

15° cut at both ends

1½" x 1½" x 7"

2 x 3 x 44⅞" L.P. to S.P.

2 x 3 x 18½"

2 x 3 x 43⅝"

Assembling the Coop

Follow these steps in order to assemble the pieces that make up the All-In-One coop. Refer back to the figures and cut list as needed.

1. **Assemble the skids and floor skirting.**

    Construct the floor skirting with the 88-inch and 44½-inch 2x4s. Fasten them together with a pair of evenly-spaced 12d nails at each joint. Place this rectangular frame on top of the 94-inch 4x4 skid runners, using the floor skirting's width as a spacing guide. Center the skirting on the runners and fasten it down with 12d nails, toe-nailing one every 12 inches or so.

    REMEMBER

    Be sure the 94-inch 4x4 skid runners are oriented so they're shorter on the bottom than the top. This allows them to slide across the ground more easily.

    Position one pair of 12-inch 2x4 risers along the skirting's short wall. The other 12-inch 2x4s should be placed so the distance from outside edge to outside edge measures 36 inches. Secure these risers to the 2x4 floor skirting with 12d nails, two per riser.

2. **Build the shelter floor.**

    Attach the 36-inch 2x4s to the tops of the risers so that they lie parallel with the 4x4 skid runners below. Use two 12d nails per riser. Next, use the 44½-inch 2x4s to complete a square floor frame for the coop shelter. Two 12d nails in each end plus another into the side of each riser will hold it fast.

    Centered between the 36-inch rim joists, insert a 41½-inch joist, and nail it in place with a pair of evenly-spaced 12d nails through each end.

    Lay the OSB sheet over the square floor frame and nail it down with 7d nails. Use one nail every 6 inches around the perimeter and across the center joist.

3. **Construct the wall frames and assemble them in order.**

    Begin with the right wall, using the figure in the earlier section "Right wall framing" for assistance. Drive two 12d nails through the top and bottom plates into each stud end. The 12¼-inch window header should be centered and positioned 7 inches under the top plate. Position the sill so the opening for the window is 12¼ inches high. Use two more 12d nails through the window's studs into each end of the window header and sill.

For the front wall, sandwich two vertical studs together where that wall meets the right wall (see the figure in the earlier section "Front wall framing"). Mark the inner stud's location on the top and bottom plates, and remove the outer stud. Fasten the inner stud to the plates with a pair of evenly-spaced 12d nails in each end. Use the 24-inch door header as a spacer to position the next stud. Use two more 12d nails in each stud end. Secure the 2x3 header on end between the door studs so that its 2½-inch-wide side faces out. Nail it fast with two 12d nails through the studs into each end. Add the lone stud at the far end of the wall with 12d nails in each end. Lastly, add the outer stud at the opposite end, sandwiched against the door's stud. Use 12d nails to secure this end stud to both plates, then drive another three 12d nails through the door stud into the outer stud.

On the left wall, use a 41-inch stud to correctly position an inner stud 1½ inches inside the top and bottom plates. Nail these studs fast with a pair of 12d nails in each end. The window header should be centered on the wall and positioned 7 inches under the top plate. The window sill should be placed to leave an opening 12¼ inches high. Use the 32¾-inch studs on either side of the window to position the 32¾-inch nest box header at the correct height. Use two 12d nails at every joint to fasten these pieces together. Finally, add the outermost wall studs and nail them in place with a pair of 12d nails in each end. Drive three more 12d nails through the inner studs and into the outer studs to create double-thick corner studs.

On the back wall frame, nail in the corner studs, along with one additional stud spaced 16 inches on-center away from one corner. Use a pair of 12d nails through the top and bottom plates into each stud end. The two remaining 41-inch studs will have an 11-inch spacer between them, creating a 12¼-inch-high opening for the chicken door. Again, use two evenly-spaced 12d nails at each joint.

Secure the right wall to the floor first, then follow with the front, left, and back walls; use 12d nails to fasten each wall section to the floor. Drive two nails through the bottom plate in between every set of studs. Nail each wall section to adjacent wall sections using 12d nails, spacing four or five evenly down each corner.

4. **Sheathe the shelter and cut out the door and window openings.**

Match the large pieces of T1-11 to the coop walls, using 39-inch-wide pieces for the left and right walls, and 45⅜-inch-wide pieces for the front and back walls. Secure each piece to its stud wall using 7d nails: one every 6 inches around the perimeter, and every 12 inches on each of the inside studs.

The interior studs show you where the various door and window openings are. Drill a hole in each corner of each opening from inside the coop. Then from the outside, connect those drilled-out holes with straight pencil lines. Carefully cut along these lines with a circular saw, jigsaw, or reciprocating saw to create the opening.

TIP

It's best to do the front wall first. You'll be able to access the interior of the shelter through the other walls' studs in order to cut out the front wall's access door opening. Then use this opening to easily get inside the coop to perform the other cuts in the wall sheathing. Saving the front wall for last means you'll be trying to squeeze through the 12-inch window, the narrow nest box opening, or the itty-bitty chicken door. (If that happens, make sure someone is rolling tape so you can become an Internet video sensation.)

5. **Secure the vertical run posts and blocking for the wire mesh.**

The 56½-inch 2x3 is set against the back wall of the coop at the end of the right wall (the one with only a window). Position the next 2x4 on that side so that it leaves an opening of 18¾ inches for the run access door. The 56½-inch 2x2 is set against the back of the coop at the end of the left wall (the one with the nest box). Horizontally-oriented 2x3s along the top and bottom of the run should be positioned to reveal 1¾ inches, enough material to allow the wire mesh to be properly secured to it later. Use two 12d nails at every joint.

6. **Attach the roof framing and assemble/install the rafters.**

Secure the 2x4 cap plate around the top of the coop and run with 12d nails every 6 inches.

Fasten the rafter 2x3s to the cap plate and each other at the peak, with the rafters spaced 18 inches apart on-center. Toe-nail once through the top front and once through the side rear of each rafter end into the cap plate using 12d nails. Use two 12d nails at each rafter peak, driving one nail from each side in a criss-cross configuration.

7. **Install the roof gables, fascia, and roof sheathing.**

Each gable is made up of two pieces of T1-11 paneling, one large and one small. The seam is concealed by the end grooves of the paneling, which meet to resemble one groove that's consistent with all the others. Be sure to install the piece that sits on the underside first, so the overlapping piece can fit over it at the seam. Use 7d nails to attach the gables to the end rafters and the cap plate, driving one nail every 6 inches or so.

Fasten the long and narrow fascia boards to the cap plate. Be sure that the bottom edge of the fascia lines up with the bottom edge of the roof gables. Drive one 7d nail through the fascia boards every 6 inches down both long sides of the coop.

Finally, secure the OSB roof sheathing. Use 7d nails spaced 6 inches apart around the outer perimeter of each sheathing panel and 12 inches apart along the inner rafters.

8. **Build the shelter access door.**

Start with the T1-11 slab. On the exterior of the slab (the grooved side), attach the 1x3 trim pieces around the door's perimeter using 7d nails every 12 inches or so. Refer to the figure in the earlier section "Shelter access door and slam strips" for a visual aid on how to orient the pieces.

WARNING

A 7d nail will fully penetrate a piece of 1x3 trim and a layer of T1-11! When you attach your trim, be judicious about where you position your nails, making sure they'll end up getting sunk in a backing stud. Otherwise, you (and your birds) will always have to be on the lookout for sharp points sticking out of the walls inside the coop.

On the interior of the door, position the 2x3 framing pieces to leave a 1-inch gap between the framing and the edge of the door slab. Drive 7d nails every 12 inches or so in each piece to secure the framing to the slab.

REMEMBER

Be sure to fasten the slam strips to the interior doorway of the coop (one on the tall side opposite the hinges and one along the top of the doorway opening) to stop the door when it's closed. Use a 7d nail every 12 inches or so.

9. **Assemble the chicken door and ramp.**

For the chicken door, attach the 1x3 trim to the outside, grooved side of the T1-11 slab with two or three 7d nails in each piece.

Using a 7d nail every 6 inches or so, fasten slam strips inside the chicken doorway's framing so that they stop the door when it's closed.

Secure the 4½-inch-long rungs to the topside of the chicken ramp with a screw at each end. The ramp can be propped up against the coop as needed or fastened permanently so that the door opens and closes freely.

10. **Build and install the nest boxes and construct the nest box door.**

Configuring the pieces of plywood as shown in the figure in the section "Nest boxes and roost bar," use 1¼-inch screws to secure the pieces to one another. Through the top of the nest box, evenly space 4 screws into each vertical divider. Use two screws through the front lip into the side of each divider.

Place the nest box on the shelter floor so that it sits in the opening on the left wall. Use 1¼-inch screws through the exterior wall to secure it, one every 6 inches or so along the top edge and sides, and one in each middle divider.

Fasten slam strips to the nest box opening with 7d nails, two evenly spaced in each side, and six across both the top and bottom.

Center the 30½-inch 2x4 on the 32½-inch-long T1-11 slab, and fasten it with four 7d nails driven through the T1-11.

11. **Install windows, all exterior trim, and doors.**

Using the manufacturer's instructions and recommended fasteners, install the window units in the openings on the coop's right and left walls.

Nail the 1x3 access doorway trim in place around the front wall opening with 7d nails, spacing the nails about 12 inches apart. Attach the access door over the opening in the shelter's front wall with a pair of T-hinges. Use 1¼-inch screws to fasten them to the wall trim, driving the screws into the stud framing behind the wall. A simple gate latch secured to both the door and the wall with 1¼-inch screws will keep the door locked when closed.

Use 7d nails every 6 inches to secure the 1x3 trim around the chicken doorway opening on the coop's back wall. Secure the chicken door to the coop's back wall trim with T-hinges and 1¼-inch screws. A gate latch installed on the door (also with 1¼-inch screws) will keep it locked closed.

TIP

We also attached the "catching" part of a second latch on the coop wall *behind* the chicken door. Positioned correctly, this second catch will latch the door in the open position as well, ensuring that your birds won't accidentally bump it closed (and lock themselves out of the shelter) while pecking about in the run.

On the right and left walls, drive a 7d nail every 6 inches or so to secure the 1x3 window trim pieces around each window.

Around the nest box opening on the coop's left wall, attach the 1x3 trim with a 7d nail every 4 inches or so. Install the nest box door to the coop wall trim using a pair of T-hinges and 1¼-inch screws. Orient the hinges on the bottom edge of the door so it opens downward.

TIP

The left wall's nest box trim includes a 5-inch 1x3 that should be secured snugly but not tightly with a single screw to the top nest box trim. Once fastened, this piece should be able to swivel up to "unlock" the nest box door. A rubber doorstop fastened to the nest box door with a single 1¼-inch screw should contact the bottom trim piece when the door is opened.

12. **Construct and install the run access door.**

Lay out the 2x3s that make up the run door as shown in the figure in the earlier section "Run access door." The piece running diagonally should fit tightly between the top and bottom pieces, and also contact the two side pieces at each end. Toe-nail all pieces together with 12d nails, two at each joint. (Alternatively, use mending plates and small screws as described in Chapter 5 to avoid toe-nailing.)

Fasten the door in place on the run's right wall (the same side as the window-only shelter wall) with a pair of T-hinges. Use 1¼-inch screws to secure the door to the center run post. Install a gate latch to the door and coop run post with 1¼-inch screws.

13. **Install drip edge on roof sheathing and roof shingles.**

Cut lengths of drip edge to the measurements of the roof sheathing on all four sides. Use a single roofing nail at every rafter, and every 6 inches along the end rafters, to install it.

Shingle the roof, starting at the bottom edge and working your way up to the peak. Keep the shingles in a straight line as you add rows, using four roofing nails per three-tab shingle. Make sure that the nailheads will be hidden by the next row of shingles. (See Chapter 8 for a more on shingling a roof.)

14. **Cut and install wire mesh in the run.**

Treat each panel of the run as its own area to be covered, cutting four large panels in all. Three narrow strips of wire mesh should be cut and installed to cover the gaps between the skids and the left, front, and right coop walls. Use fencing staples to install the wire mesh, spaced about 2 or 3 inches apart around the entire perimeter of each piece.

15. **Install the roost bar and tow chains.**

The roost bar should fit between two opposite studs on the left and right walls and be screwed into place (use 1¼-inch screws, 2 toe-screwed through each end).

Fasten the tow chains to the coop's skids with lag screws. Use one chain on the front wall and one chain on the back wall.

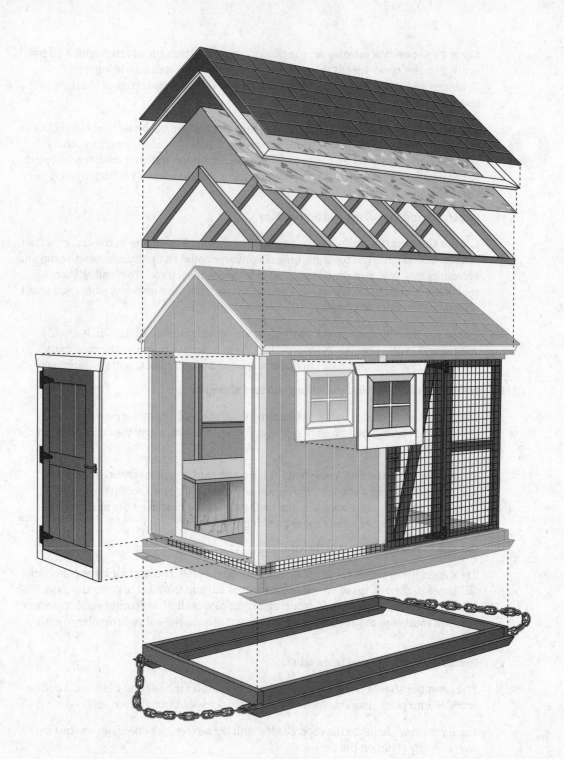

Chapter **16**
The Walk-In

For the gung-ho owner who wants to start right out of the box with a lot of birds or the veteran caretaker who's finally ready to supersize the flock, a large coop is required. Most of these folks want loads of square footage for their hens, but also plenty of room for supplies, tools, and themselves. This sort of structure usually ends up looking like a toolshed. In the chicken biz, it's called a *walk-in coop.* And while you can seek out plans for an actual toolshed and retrofit it to your flock's needs, why not start out with a shelter that was specially-designed for chickens to begin with? (See Chapter 2 for more on various coop styles and to see how a walk-in compares to smaller designs.)

Vital Stats

» **Size:** Our largest coop, the Walk-In, has a footprint that measures 8 feet x 8 feet. At the peak of its gabled roof, it stands 9 feet tall.

» **Capacity:** Holy henhouse! The Walk-In's 64 square feet easily accommodate up to 30 chickens, making it ideal for even the most ambitious backyard chicken-keeper.

» **Access:** Provided the caretaker isn't also a member of the Boston Celtics, entry into the coop is pretty easy with a door that's 6 feet tall. Chickens (of any size) have their own access door.

» **Nest boxes:** If you're sheltering close to 30 birds, you'll need a veritable chicken condo when it comes to nest boxes. The Walk-In's design includes a bank of ten nest boxes, accessed by the caretaker through an exterior hatch.

» **Run:** The Walk-In is a freestanding structure that does not have an incorporated run. It's meant for areas where the birds will be allowed to free range or where a separate run will be built, as described in Chapter 10.

» **Degree of difficulty:** "Big" doesn't have to mean "difficult." We've designed our largest coop with the *For Dummies* reader in mind. If you're comfortable with the skills and techniques in Chapters 5 through 10, you'll be able to construct this coop with little problem.

» **Cost:** Even with the extra materials needed for a coop this size, we bought everything we needed to build the Walk-In for around $1,000.

Materials List

Here's our shopping list for the Walk–In. If you need to, you can change the materials to better suit your needs, budget, or capabilities. (Chapter 4 offers help in wading through some of the more common alternative materials.)

Lumber		Hardware		Fasteners	
7	4' x 8' sheets of ⅝" OSB	1	18" x 27" window	400	12d nails (or 3" wood screws)
11	4' x 8' sheets of T1-11	4	10' lengths of drip edge	550	7d nails (or 1¼" wood screws)
2	8' lengths of pressure-treated 4x4 lumber	3	bundles of roof shingles	300	1" galvanized roofing nails
73	8' lengths of 2x4 lumber	2	14¾" x 4¾" gable vents	100	1¼" wood screws
4	8' lengths of pressure-treated 2x4 lumber	4	8' lengths of metal corner trim		
15	8' lengths of 2x3 lumber	7	5" T-style hinges		
10	8' lengths of 1x3 lumber	3	barrel bolts		
2	8' lengths of 1x2 lumber	1	gate latch		
		1	doorstop		

Cut List

The following sections break down the Walk-In coop into smaller, more manageable pieces. We detail how to cut your lumber so that you can assemble each piece, one at a time, and then put the pieces together at the end. (Chapter 5 reviews making safe and accurate lumber cuts.)

This coop features several pieces that are cut on an angle. Some boards in the cut list specify L.P. or S.P.:

>> *L.P.* stands for "long point," where the board's listed length refers to the tip-to-tip measurement at the board's longest point, with the angle making one side shorter.

>> *S.P.* is an abbreviation for "short point." When you see it, know that the board's listed length measures just to the end's shortest point, meaning that the long end of the board will be greater than the cut list's specified length.

   For example, a 2x4 listed as 36 inches S.P. with a 45-degree angle will actually be almost 39½ inches long at its longest point. (See Chapter 5 for details on how to lay out angles before cutting to ensure accuracy.)

This coop features several narrow pieces of lumber that must be *ripped*, or carefully cut lengthwise from wider lengths of stock lumber. In addition, some wide sheet goods will need to be ripped into narrower strips of wood. Refer to Chapter 5 for more on how to safely and accurately perform a rip cut.

Skids and floor

Quantity	Material	Dimensions	Notes
1	Sheet of OSB	48" x 96"	
1	Sheet of OSB	44" x 96"	
2	Pressure-treated 4x4s	96" long	
2	Pressure-treated 2x4s	96" long	
2	Pressure-treated 2x4s	89" long	
5	2x4s	89" long	

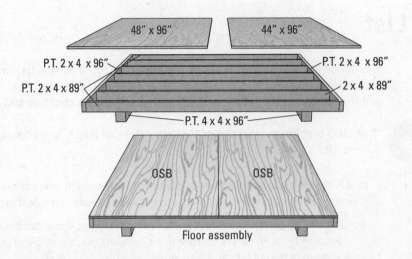

Floor assembly

Front wall framing

Quantity	Material	Dimensions	Notes
2	2x4s	96" long	
1	2x4	89" long	
7	2x4s	74" long	
2	2x4s	69" long	
3	2x4s	33⅞" long	
2	2x4s	18¼" long	

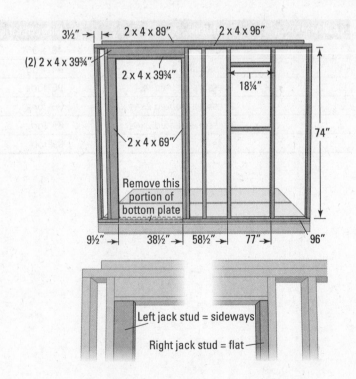

Back wall framing

Quantity	Material	Dimensions	Notes
2	2x4s	96" long	
1	2x4	89" long	
7	2x4s	74" long	

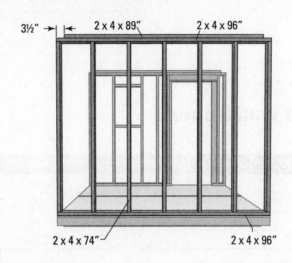

Right wall framing

Quantity	Material	Dimensions	Notes
1	2x4	92" long	
2	2x4s	85" long	
2	2x4s	74" long	
4	2x4s	63¼" long	
2	2x4s	48¼" long	
5	2x4s	34¼" long	
4	2x4s	24¼" long	
5	2x4s	9" long	

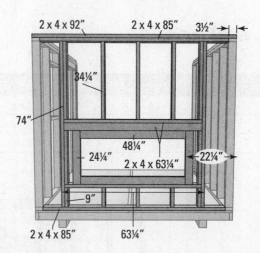

Left wall framing

Quantity	Material	Dimensions	Notes
1	2x4	92" long	
2	2x4s	85" long	
5	2x4s	74" long	
1	2x4	14½" long	

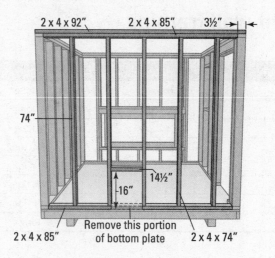

Front wall sheathing and trim

Quantity	Material	Dimensions	Notes
1	Sheet of T1-11	48" wide x 80¼" high	
1	Sheet of T1-11	30¾" wide x 6" high	
2	Sheets of T1-11s	9" wide x 80¼" high	
2	1x3s	75" long	
1	1x3	37" long L.P.	Both ends trimmed at a 15-degree angle

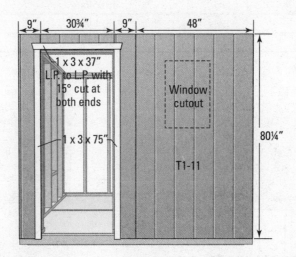

Back wall sheathing

Quantity	Material	Dimensions	Notes
2	Sheets of T1-11	48" wide x 80¼" high	

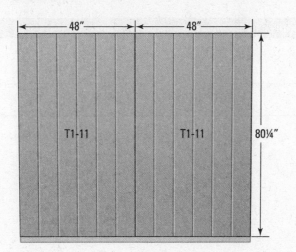

Right and left wall sheathing and trim

Quantity	Material	Dimensions	Notes
2	Sheets of T1-11	48" wide x 80¼" high	
2	Sheets of T1-11	44" wide x 80¼" high	
1	1x3	55¼" long L.P. to L.P.	Both ends trimmed at a 15-degree angle
1	1x3	53½" long	
1	1x3	21¼" long L.P. to L.P.	Both ends trimmed at a 15-degree angle
1	1x3	19½" long	
2	1x3s	16½" long	
2	1x3s	16" long	

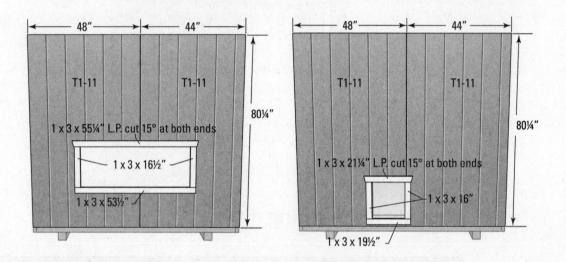

Roof rafters

Quantity	Material	Dimensions	Notes
12	Sheets of OSB	16" x 6"	Make a 22.5-degree cut starting at the center point on both ends
14	2x4s	53¼" long L.P. to S.P.	Both ends cut at a 22.5-degree angle

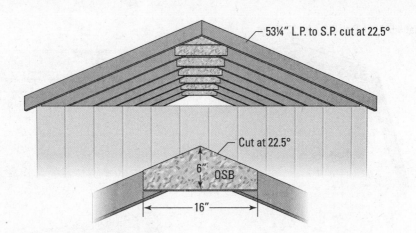

53¼" L.P. to S.P. cut at 22.5°

Cut at 22.5°

6" OSB

16"

Gable ends and soffits

Quantity	Material	Dimensions	Notes
2	Sheets of T1-11	48" wide x 25" high	
4	Sheets of T1-11	48" wide x 16" high	
2	Sheets of OSB	96" x 2⅞"	
2	1x2s	96" long	

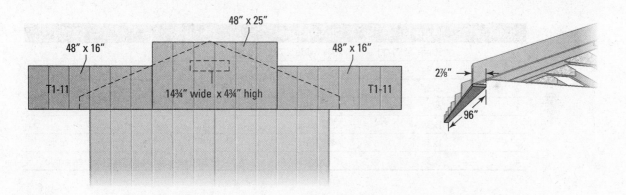

48" x 25"

48" x 16"

48" x 16"

T1-11

14¾" wide x 4¾" high

T1-11

2⅞"

96"

Roof, gable overhangs, and fascia

Quantity	Material	Dimensions	Notes
2	Sheets of OSB	48" x 84"	
2	Sheets of OSB	48" x 20"	
2	Sheets of T1-11	84" wide x 5½" high	
2	Sheets of T1-11	20" wide x 5½" high	
4	2x4s	53⅜" long L.P. to S.P.	Oriented to sit flat; both ends cut at a 22.5-degree angle

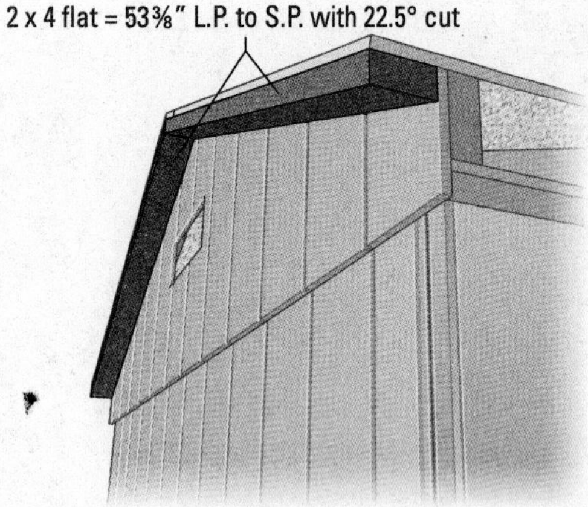
2 x 4 flat = 53⅜″ L.P. to S.P. with 22.5° cut

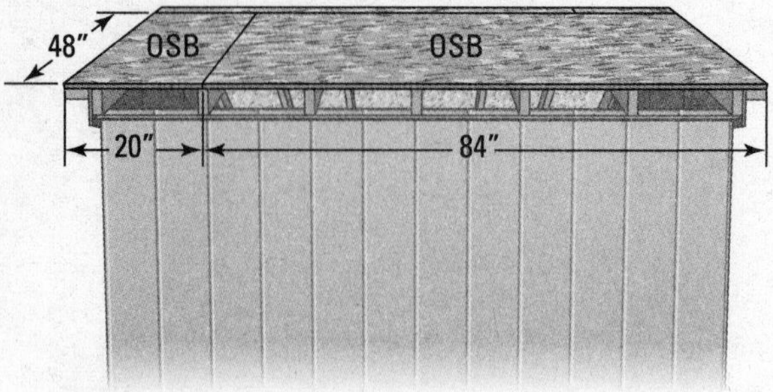
48″
OSB
OSB
20″
84″

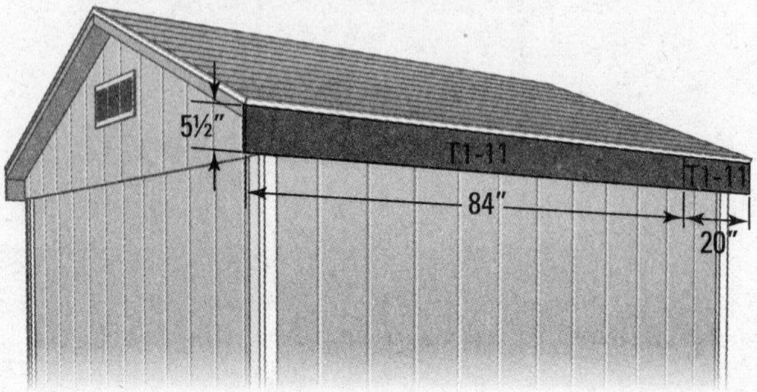
5½″
T1-11
T1-11
84″
20″

Access door

Quantity	Material	Dimensions	Notes
1	Sheet of T1-11	29¾″ wide x 72½″ high	
3	2x3s	65″ long	
2	2x3s	27¼″ long	
1	2x3	20¾″ long	
2	1x3s	72½″ long	
3	1x3s	24¾″ long	
4	1x3s	8″ long L.P.	Both ends cut at a 45-degree angle

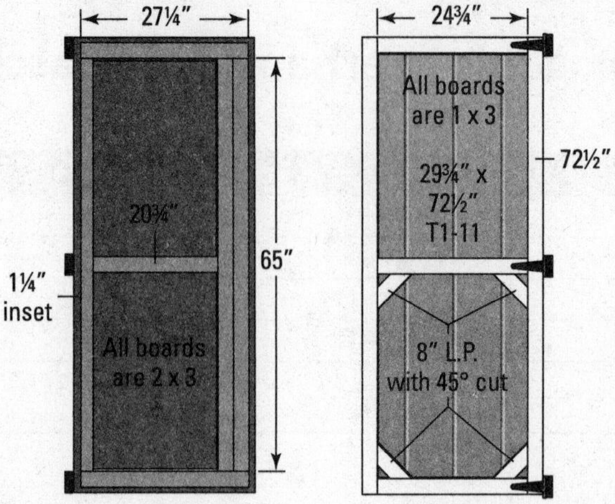
27¼″
20¾″
65″
1¼″
inset
All boards
are 2 x 3
24¾″
All boards
are 1 x 3
72½″
29¾″ x
72½″
T1-11
8″ L.P.
with 45° cut

Chicken door

Quantity	Material	Dimensions	Notes
1	Sheet of T1-11	14" wide x 15½" high	
2	1x3s	14" long	
2	1x3s	10½" long	

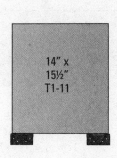
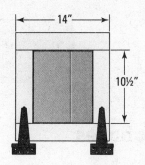

Nest box door

Quantity	Material	Dimensions	Notes
1	Sheet of T1-11	47¾" wide x 16¼" high	
2	2x3s	47¼" long	
2	2x3s	11" long	
2	1x3s	47¾" long	
2	1x3s	11" long	
4	1x3s	7" long L.P.	Both ends cut at a 45-degree angle

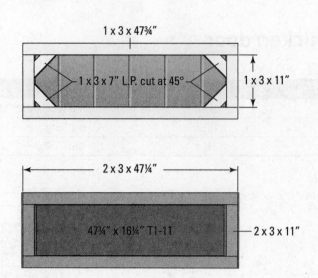

1 x 3 x 47¾"

1 x 3 x 7" L.P. cut at 45°

1 x 3 x 11"

2 x 3 x 47¼"

47¾" x 16¼" T1-11

2 x 3 x 11"

Nest boxes

Quantity	Material	Dimensions	Notes
3	Sheets of OSB	58¾" x 11⅛"	
2	Sheets of OSB	24" x 11⅛"	
8	Sheets of OSB	11⅛" x 11⅛"	
2	2x3s	63" long	
2	2x3s	20" long	
2	2x3s	14" long	
4	1x3s	60" long	
2	1x3s	51½" long	
2	1x3s	15½" long	

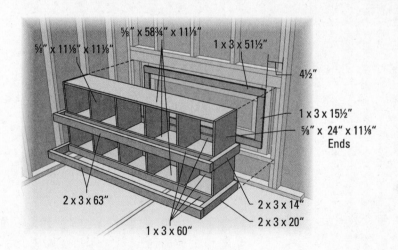

⅝" x 58¾" x 11⅛"

⅝" x 11⅛" x 11⅛"

1 x 3 x 51½"

4½"

1 x 3 x 15½"

⅝" x 24" x 11⅛"
Ends

2 x 3 x 63"

2 x 3 x 14"

2 x 3 x 20"

1 x 3 x 60"

Roost

Quantity	Material	Dimensions	Notes
2	2x4s	27⅞" long L.P. to S.P.	Both ends cut at a 25-degree angle
5	2x3s	58⅝" long	
2	2x3s	30¼" long L.P.	One end trimmed at a 25-degree angle
2	2x3s	20¼" long	
2	2x3s	19¾" long L.P.	One end trimmed at a 25-degree angle

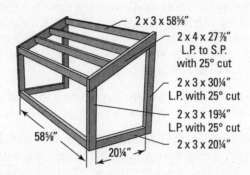

2 x 3 x 58⅝"
2 x 4 x 27⅞" L.P. to S.P. with 25° cut
2 x 3 x 30¼" L.P. with 25° cut
2 x 3 x 19¾" L.P. with 25° cut
2 x 3 x 20¼"
58⅝"
20¼"

Assembling the Coop

Here are the steps to follow to assemble the Walk-In coop. Refer back to the figures and cut list as needed.

1. **Assemble the floor and put it on the skids.**

   Start by building the subfloor frame. Use the 8-foot pressure-treated 2x4s for the rim joists of the frame, and use 16-inch on-center spacing for the 89-inch floor joists. (The two pressure-treated joists are the end joists.) Drive two evenly-spaced 12d nails through the rim joist into each joist end.

   Next, position the 8-foot 4x4 skids underneath so that each sits 10¾ inches inside the subfloor frame. Before attaching the OSB floor, secure the subfloor joists to the skids by toe-nailing into each skid with two 12d nails through each joist, one from either side. (On the end joists, both nails will have to be driven from the outside of the subfloor frame, to keep the nails from protruding through the skid ends.)

Finally, attach the pieces of OSB on top of the joists with 7d nails, using one every 6 inches across each joist.

2. **Build the wall sections in order: front, back, right, left.**

Use 16-inch on-center spacing for the studs on all four walls.

On the front wall, place the doorway's left jack stud (from outside the coop) first. It should be set sideways, so the 3½-inch face faces out and the measurement from the outside of the corner stud to the inside of the sideways jack stud is 9½ inches. (The right jack stud is oriented normally, 38½ inches from the outside corner stud.) For the door header, use three 2x4s: two sandwiched together and stacked on top of one lying flat. The window jack studs should be set 58½ inches and 77 inches from the corner stud. Position the window header so that it starts 10½ inches down from the top plate and creates an opening that's 27¼ inches tall. (The sill of the window is set 39½ inches up from the floor; the header is set 70½ inches off the floor.) Make all attachments with 12d nails, two in each stud end.

Finally, position the 89-inch cap plate atop the wall to leave a 3½-inch gap on either side. Nail it to the top plate with 12d nails, two at every other stud.

REMEMBER

Because you'll cut the window opening in the exterior paneling after it's nailed to the frame, you can set the window at any height that's convenient for you. Just verify that the *rough opening* (the hole in the wall for the window) is slightly bigger than your window. For our 18-x-27-inch window, the opening is 18¼ inches wide x 27¼ inches high to allow the window unit to slide into place.

After assembling the back wall's top and bottom plates to the studs with a pair of 12d nails in each end, fasten the cap plate to leave a 3½-inch gap on either end to allow for the cap plates of the right and left walls. Use two 12d nails at every other stud location.

The right wall features the nest box opening, which spans the distance between the five inner studs. Use the 9-inch cripples as spacers to position it at the correct height on the wall. The opening's header comprises three 2x4s, oriented like the doorway header on the front wall, as described earlier in this step. Inside the main nest box opening is a smaller "frame" opening, created with 2x4s turned sideways. This opening should be centered on the wall so that the measurement from the outside corner stud to the inside of this inner smaller opening is 22¼ inches. Make all connections with a pair of evenly-spaced 12d nails in each stud end. Where studs are doubled up, nail them to one another with three or four 12d nails. Where the nest box header meets the nest box frame, toe-nailing will be required. Secure the cap plate so that it overhangs the top plate by 3½ inches on either side.

The chicken door is located on the left wall. The 14½-inch header that serves as the top of this doorway should be placed to leave an opening that's 16 inches high once the portion of the bottom plate in the doorway is removed. (That puts it 14½ inches up from the bottom plate during framing.) Use two 12d nails in each stud end to fasten everything together. Attach the cap plate with 3½ inches of overhang on either end of the top plate.

As you raise each wall into a standing position, use two 12d nails in between every set of wall studs to secure the wall to the subfloor. Use a pair of 12d nails on the right and left walls where the cap plates overlap the top plates of the adjoining front and back walls.

3. **Cut and fasten the exterior wall sheathing and trim.**

REMEMBER

Set each piece of exterior sheathing ½-inch down from the top of the cap plate. This allows for the overhang of the roof rafters and will be covered later.

The front wall uses one full sheet over the half of the wall with the window. (You may cut the window out any time after securing the sheathing in place; see Chapter 8 for tips on doing this.) When cutting the smaller pieces to fit around the doorway, pay attention to the grooves in the sheet lumber so that they match up with those in the full sheet and result in consistently-spaced grooves across the entire front wall. Drive 7d nails through the T1-11 into the studs, using one nail every 6 inches around the perimeter; every 12 inches on inner studs.

The back wall is easiest, using two full 4-foot-wide sheets butted together. Use a 7d nail every 6 inches around the perimeter of the sheathing, then nail it once every 12 inches to the inner studs.

Attach the right and left wall sheathing using a full 48-inch-wide sheet in conjunction with a sheet ripped to 44 inches wide. Fasten the sheets to the walls with 7d nails, one every 6 inches around the perimeter, and one every 12 inches on the inner studs. Carefully cut out the openings for the nest boxes and chicken door.

TIP

Position the ripped edge of the 44-inch-wide sheet on the outside corner of the coop wall. This puts the factory edge against the full sheet, maintaining the spacing of the grooves. The ripped edge will be covered by corner trim.

Attach exterior trim pieces around the openings for the access door on the front wall, the chicken door on the left wall, and the nest box hatch on the right wall; use 7d nails spaced 6 inches apart on all trim pieces. Fasten metal corner trim to all four outside corners of the coop using 7d nails and/or construction adhesive.

4. **Assemble and set the roof rafters.**

Use *gussets* — small triangular pieces made of OSB — at the rafter peaks to hold them together. Each interior rafter should use two gussets, while the end rafters should each employ only one, set on the inside of the coop. (Gable sheathing will help strengthen the outer connection of the end rafters.) Assemble the gussets and rafters using six to eight 7d nails spaced evenly around the gusset and penetrating the rafters.

Set the rafters on top of the walls using 16-inch on-center spacing. Toe-nail the rafters to the cap plates with 12d nails, two at each rafter-to-plate intersection.

5. **Cover the gables and make rafter soffits.**

TIP

Use a chalk line to snap a mark 1¼ inches down from the top of the cap plate around the entire coop. Set pieces of sheathing on this line as you cover the gables. (Flip to Chapter 5 for tips on using a chalk line effectively.)

Start with a 25-inch-tall piece of exterior sheathing, using the grooves of the main wall to help you position it at the peak. Attach this piece with 7d nails driven every

6 inches into the end rafter and cap plate; then do the same on either side with a 16-inch-tall piece. After fastening, cut away the excess sheathing as shown in the figure in the earlier section "Gable ends and soffits."

Position the 8-foot 1x2s to the front and back walls at the chalk line. The bottoms of the 1x2s should sit perfectly level with the bottoms of the rafter tails; if they don't, adjust the 1x2s until they do. Nail them to the coop walls with a 7d nail at every stud location. Then attach the thin 2⅞-inch soffit strips of OSB to the bottoms of the rafters and the 1x2 strips, using a 7d nail in each rafter tail and another at each stud location along the 1x2.

Position your vent at the desired location in each gable, cut the hole, and secure the grate with the small screws included.

6. **Finish the roof.**

Secure the roof sheathing to the tops of the rafters, placing a 7-foot piece and a 20-inch piece on either side and leaving an equal amount of overhang on either side; use a 7d nail every 6 inches where the edges of the OSB sit on a rafter, and every 12 inches along the inner rafters. Fasten the flat 2x4 overhang pieces (the ones that are 53⅜ inches) to the sheathing's underside at either end. Use 7d nails driven at a slight angle through the sheathing and into either 2x4 to keep the nail points embedded in the 2x4; go with one nail every 12 inches or so.

Use thin 5½-inch-wide strips of exterior paneling as fascia boards to close off the openings between rafters. Orient the 20-inch piece and the 7-foot piece to create a seamless run of consistently-spaced grooves. Use a pair of evenly-spaced 7d nails in each rafter end for this step.

Attach drip edge to the roof sheathing with roofing nails (one in each rafter on the long sides and every 12 inches along the short sides), and shingle the roof with roofing nails (see Chapter 8 for the how-to on these tasks).

7. **Build and install your doors.**

Position the interior blocking pieces of the access door so that they're 1¼ inches in from the outer edges of the door itself. Double up two vertical 2x3s on the non-hinged side, with the outer one placed flat and the inner one turned sideways. (This adds strength and helps serve as a pull handle from inside the coop.) Center the middle horizontal piece on the door, for both the interior blocking and the exterior trim. Use your larger 12d framing nails to assemble the door frame, placing two in each end at every joint, and three or four along the two doubled-up 2x3s. Secure the frame to the T1-11 slab with 7d nails driven every 12 inches.

Attach the 1x3 exterior trim along the door slab's outer edges as shown in the figure in the "Access door" section, placing a 7d nail every 12 inches.

REMEMBER

The short angled pieces that help make up the exterior trim package of the access door (and the nest box hatch a little later) may look like decorative accents only, but they help keep the doors square. They're not optional.

The chicken door features exterior trim on a small T1-11 slab, but no interior blocking. Fasten the 1x3 trim to the grooved side of the T1-11 with a 7d nail at each end of each piece.

Secure the 2x3 interior blocking to the backside of the nest box door slab with 7d nails every 12 inches. Position the 1x3 exterior trim on the other side and fasten the pieces with 7d nails spaced about 10 inches apart (and at either end of the short angled pieces).

Add hinges (two for the nest box hatch, two for the chicken door, and three for the access door) and secure them to their respective coop walls with 1¼-inch screws. (In our coop, the access door hinges are installed on the right side, and the hinges for the chicken door and nest box door are installed along the bottom edge so that they open downward.) Barrel bolts and gate latches may also be installed at this time: a gate latch at the access door, one barrel bolt at the chicken door, and a barrel bolt on either end of the nest box door. A rubber doorstop can be mounted to the outside of the nest box hatch to protect it from banging into the coop wall when opened. Use 1¼-inch screws with all of these pieces of hardware.

8. **Construct the nest boxes.**

Build the ten-stall nest box as a freestanding unit first (refer to the figure in the "Nest boxes" section for a visual), using 1¼-inch screws. Attach the 1x3 blocking with 7d nails to the inside of the nest box wall opening (placing the vertical pieces 4½ inches in from the wall studs), and use it to position the nest box in place. The 2x3 perches can be used to help secure the nest box unit to the coop wall; attach them to one another with two 12d nails in each end. The side 2x3s should fit snugly against the wall studs of the coop; make the connection with two evenly-spaced 12d nails driven through the stud into the perch.

9. **Build the roost.**

The front roost bar sits 28 inches away from the back bar. Position the middle two roost bars evenly between the front and back, with about 9 to 10 inches between them. The roost may be moved around inside the coop, or fastened in place by toe-nailing it to the coop floor with 12d nails. (Most coop-owners place it along the back wall; see Chapter 9 for ideas on where the roost should ideally be located.)

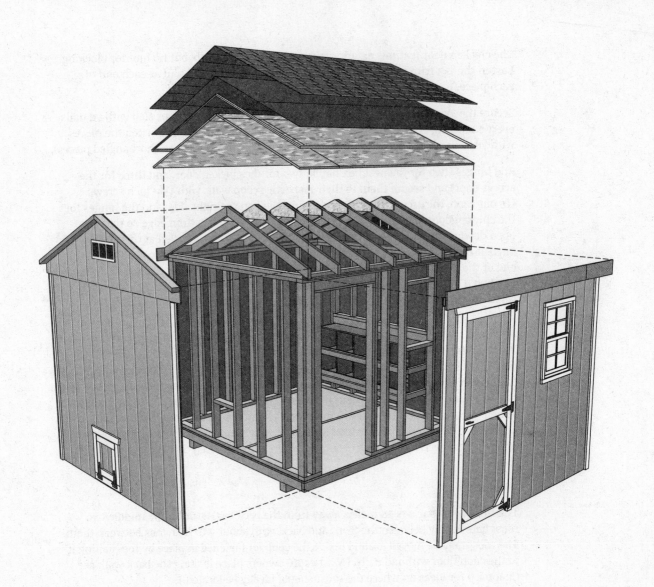

The Part of Tens

4

IN THIS PART . . .

Everybody loves a good Top Ten list. In this part, we present two of them.

The first runs down some common things (we won't call them "mistakes") that folks who have already built a chicken coop would do differently if they could do it all over.

The second list is sort of a fantasy coop wish list — ways you could amp up your coop with a few bells and whistles that might make your job as a chicken caretaker easier, cleaner, or just more fun.

» **Making the coop convenient and accessible for you**

» **Easing the headaches of cleaning and long-term maintenance**

» **Building a coop that will last the first time around**

» **Outsmarting predators who dig**

Chapter **17**

Ten or So Things Novice Coop-Builders Would Have Done Differently

Scientists have calculated that the average length of time between the moment a DIYer finishes building a chicken coop and the moment that second-guesses and mild regret kick in is approximately eight seconds.

Okay, not really. But it is safe to say that almost anyone who has ever built something with his own hands, no matter how thoughtfully designed, meticulously planned, and expertly crafted that something is, pretty quickly comes up with at least one or two ways that it could be even better. Compiling a "coulda, shoulda, woulda" list when all is said and done is practically its own step in any building project.

While your chickens will no doubt love their new home and find it to be perfect in every way, you'll almost assuredly start a mental rundown of the things you would have done differently if you had it to do over again — and will do differently when you build your *next* coop. (And trust us, for most chicken folk, there's *always* a "next coop.")

So what did the coop-builders who walked in your work boots before you learn during their first DIY project? Read on, because this chapter clues you in to the ten (or so) most common things they'd do differently on a second coop build. Maybe they're tips you can use on your first coop build.

Make the Coop Bigger

Ask 100 people who have built their own coop what they would have done differently, and more of them will shout out these three words than any other answer: "Make it bigger!" Almost no one ever says, "Gee, I wish I'd made my coop smaller."

Judging your coop's size while it's still on paper is sort of like looking at it in a rearview mirror: Objects may appear bigger than they are. And once you build a coop, it tends to get a lot smaller in a big hurry. When you finally move your birds in, you just may find yourself wondering what went wrong and where the rest of your coop went.

A bigger coop means several things, and almost all of them are good, including the following:

>> **It means more room for the chickens you have.** This translates to happier chickens who aren't crowded, squabble less, don't get sick as often, and make your life as their caretaker much more pleasant.

>> **It means more room for you.** You'll find this aids considerably in cleaning, feeding, egg-gathering, and storing supplies.

>> **It means more room for new chickens!** Keeping a backyard flock is a fantastic hobby that usually starts small and grows exponentially, often turning into a full-blown obsession. It would be a shame to have to curtail your exciting new venture simply because you didn't plan ahead when building your coop and don't have room for more birds.

REMEMBER

It almost never hurts to build your coop larger than you think it needs to be. If you find you want to increase the size of your flock, it's a matter of bringing some new birds home. If you never get past the starter-flock stage, at least the birds you do have will be able to spread out and live a comfy, spoiled life.

Make the Coop Taller

"Taller" is different than "bigger." A "bigger" coop implies more square footage — more floor space for chickens. A "taller" coop, on the other hand, simply means more headroom for you. One of the most common coop-building regrets is creating a structure that doesn't accommodate the caretaker. Your hens will call a 5-foot-tall ceiling "vaulted"; you'll use more colorful adjectives the first time you whack your noggin on it. And kick yourself with every subsequent bump for not just going up a foot or two.

Consider the Location More Carefully

The perfect spot in the yard for your new coop can be a tricky call. You want it a little ways away from the house — but not too far. You'll be making that trek often. At all hours. And in all kinds of weather conditions. Just like the mailman, "neither snow nor rain nor heat nor gloom of night. . . ."

The point is to make sure that walk is one you won't mind (too much) making a few times a day. Sadly, a lot of flocks end up being neglected because an eager new caretaker puts his sparkly new coop waaaay out in the back corner of the yard and then can't find the time to go perform his daily duties.

Also, check the spot you're considering at different times during the day. What's the sun exposure like? The way the sun breaks over a nearby treeline has determined which direction many a coop has faced. And many more owners wish they would have bothered to check this out before building a coop with all its windows facing north.

REMEMBER

If you're constructing a tractor-style coop that will be moved around (the ones in Chapters 14 and 15 even have tow chains!), you'll find a few sweet spots that work well for you and your birds. But for a stationary coop, think long and hard about exactly where that coop should go before you build. Flip to Chapter 2 for more guidance on choosing your coop's location.

Don't Cheap Out on Materials

As with most hobbies, taking up the raising of backyard chickens can mean an initial outlay of a fair bit of money. (Unlike golf, though, this hobby at least tries to pay you back, in fresh eggs or meat.) And just a few trips to the building supply center to buy materials to build your own coop from scratch can put a decent-sized dent in the ol' credit card statement. The costs of exterior-grade lumber, weather-resistant fasteners, and heavy-duty fencing can add up in a hurry. You may be tempted, then, to cut corners here and there in order to bring the overall price tag down.

Online chicken forums, though, are packed with stories that go, "I didn't want to fork out for X (insert high-quality building material here), so I tried to get by with Y (insert flimsy, cheap alternative here). Now I have to Z (insert messy/unpleasant/time-consuming/even-more-expensive task here)."

REMEMBER

You've heard the saying, "You get what you pay for." It's a famous saying because it's true. Pay for materials that will last a long time right up front, and you'll never have to worry about them again. Try to save a buck by using particle board, nails that rust, and thin chicken wire, and you're practically guaranteed to end up replacing some or all of them. And maybe your chickens, too. (Chapter 4 has more on high-quality materials we recommend.)

Use Screws Rather than Nails

This tip may stir up some grumblings from die-hard DIYers who swear by nails for their high speed and low cost. But if you make a mistake during the build and need to reposition or completely relocate a board you've already fastened in place, would you rather wrestle with the claw end of a hammer, trying to get enough leverage to pry the board up without destroying it in the process so you can reuse it, or switch your drill to the "reverse" setting and back out that baby in a matter of seconds? We thought so.

But using screws can also pay off down the road, if you ever decide to build on to the coop or simply change some things around. One caretaker found she wanted to mount the nest boxes on the opposite wall of her shelter. Another decided that the roost bar was just too high for his fat little cluckers to reach. Someone else considered moving the attached run from the left side of the coop to the right side. If they had used screws instead of nails to make those attachments in the first place, they could have knocked these things off their to-do list during a commercial break of *American Idol*. (Head to Chapter 4 for more information on the differences between screws and nails.)

Elevate the Coop Off the Ground

Lots of rookie DIYers build their coops to sit on the ground because, well, it's easy. That's where the chickens are, right? But plenty of folks realize after they've had a coop for a while that getting the birds' shelter off the ground, even by just a few inches, can sometimes make a world of difference.

First, think about how wet the ground is a lot of the time. A shelter sitting on the ground gets a lot wetter than one built on legs. Wood rots more quickly sitting on the ground. Building your coop so that it sits on concrete footers and posts (as described in Chapter 6) or even on concrete blocks can keep all that excess wetness well away from your henhouse.

Also, consider your maintenance. Do you really want a coop that you have to stoop or squat to reach into? Putting the coop on a more comfortable level makes cleaning a little less backbreaking, egg-collecting a little quicker, and scooping up chickens a little easier.

 Trust us; your pullets won't mind the penthouse treatment. Just be sure to build them a simple
TIP   ramp or steps, like we describe in Chapter 9.

Make the Doors Wider

Picture yourself going into your coop. Now picture yourself trying to squeeze through that doorway while carrying a flapping bird. Make the doorway wider than you think it needs to be. You'll be going in and out with tools, supplies, pails of water, bags of feed, and even the occasional wriggling chicken. Build a doorway that will accommodate those tasks. (One builder of a walk-in coop used his wheelbarrow as his guide in determining the doorway's width.) We introduce the basics of building coop doors in Chapter 8.

Consider How to Clean the Coop

What's the best part of keeping chickens for most caretakers? Watching the birds' crazy antics? The self-sustainability? The companionship of a whole flock that depends on you for survival? The farm-fresh eggs? The answer may differ depending on whom you ask, but rest assured that no one ever claims that "cleaning the coop" is her favorite thing about having a backyard flock.

There's no way to sugar-coat it: Chicken coops get filthy. And one of your jobs as caretaker is to keep things clean (at least somewhat), for them as well as for yourself. Some forethought in the design phase of the build can make that regular maintenance a lot easier down the road.

Nowhere in the coop does a little pre-planning pay off more than around the roost, where your birds will perch and sleep for the night. The area directly underneath the roost becomes one big "drop zone," where the lion's share of your chickens' waste will pile up. So it's not a good spot for the flock's open nest boxes. Or their feed and water. Or your access door.

Make sure that corner of the coop is easy for you to reach with whatever tools you think you may need. Underneath the usual bedding or shavings, some owners opt to put down a material like linoleum that can be lifted up, taken out, and hosed off. Some use litter boxes. Some leave wire-mesh-covered openings in the floor for the waste to fall through. Others build trays that can be pulled out from outside the coop for the ultimate in no-mess cleaning (more on that in Chapter 18).

Whatever method you use, it's easier to *build* an easy-to-clean coop than it is to *make* a coop easy-to-clean, so think about it up front. Flip to Chapter 9 for more information on dealing with a roost's drop zone.

Make the Nest Box Easily Accessible

Anyone who has ever had to crawl on his hands and knees into a dirty, stinky chicken coop is nodding his head vigorously at this piece of advice. Having to scoop chickens out of the way or hurry one out of a nest box just to get at the day's eggs is messy and unpleasant (and a good way to break a few eggs in the process). Consider your coop design carefully to ensure that you'll have easy access to your nest boxes later on.

Many caretakers build their nest boxes so that they stick out of an exterior wall of the coop and feature a latching lid on top. (The plans for several of our coops in Part 3 use this approach.) This simple technique solves two common coop problems:

>> It frees up a little more floor space inside the coop for your birds.

>> It allows you to collect eggs without ever setting foot (or hands and knees) inside the coop itself.

TIP

If your coop design can't accommodate an exterior nest box, you may still be able to construct a hatch in an exterior wall that you can open from the outside to access the nest boxes. Follow the steps for building a basic shed door in Chapter 8, only make it smaller. And for the fundamentals of building nest boxes, flip to Chapter 9.

Paint the Coop Before Assembly

Many (if not most) folks who build a chicken coop do so with a cardboard box full of chicks in the garage anxiously awaiting their new home. Time is often a factor, and picking a design, planning the build, making a shopping list, buying the materials, prepping the site, cutting the lumber, assembling the coop, and doing the finish work is often enough of an exercise for many DIYers without throwing another step in there that could be done later.

But whole flocks of chicken-owners have realized that moving all of their birds out of the coop and completely cleaning it out from top to bottom just to spend a weekend painting the darn thing is no fun at all. And don't even get them started on what to do with their hens while the paint dries.

A lot of coop-builders have had great success with executing the cut list and building each piece of the coop (like in our plans in Part 3) and then painting the pieces before final assembly. This way, you assemble painted pieces to construct the coop, can move the birds in immediately upon completion, and can do any touch-up work at your leisure without disturbing your hens.

WARNING

Pressure-treated lumber needs to dry completely before it can be painted because it's still "wet" with chemical preservatives when you buy it. That means you'll need to let it sit exposed to the elements for up to six months!

Reinforce the Run Underground

You know those prison movies where the bumbling guards come around the corner one morning and choke on their donuts when they find a big fat hole in the yard, where several inmates went "under the wire" in the middle of the night and are now gone forever? That same dumbfounded look can be seen on the faces of countless chicken owners who have underestimated the digging capabilities of hungry predators, and thus find themselves the proud owners of empty chicken runs.

Most chicken runs take a simple 2x4 frame and wrap it in wire mesh. But all too often, that "cage" simply rests on open ground. That's fine for keeping birds in, but sometimes it's not so effective at keeping determined raccoons out. If a predator who has a taste for chicken can't get through the wire or over the top of the fencing, he'll often start burrowing under the run. And, just like those big-screen cons serving life terms in Cell Block D, that furry little chicken-napper's got nothing but time to work on his tunnel. One morning, you might wake up to an empty run, with only a few fluttering feathers to serve as a reminder that you could have just extended the wire mesh, burying it a foot or two underground, and prevented a subterranean snack attack. Chapter 10 has the scoop on reinforcing your run underground.

Chapter **18**

Ten or So Cool Ideas to Trick Out Your Coop

For many chicken-owners, a coop is never really done. There's always some new brainstorm of an idea to incorporate, some spiffy new gadget to install, some clever trick you saw somewhere that just might work on your own coop. True to their can-do attitudes and self-sustainable natures, most coop-builders are always on the lookout for ways to tweak their coop to make the chicken-raising adventure easier, cleaner, or just more fun.

Most folks with a backyard chicken coop also keep a running mental list of things they'd include in a "dream coop." These things go beyond what they would have done differently, which we tackle in Chapter 17. These are the little extras that, while certainly way above and beyond the reach of most first-time coop-builders, veteran caretakers have put on their all-time wish lists for that "ultimate" chicken coop they all want to build someday.

This chapter includes ten or so of these "extras." Some are pie-in-the-sky, luxury-type items that are nice to dream about, but perhaps not terribly practical for the average DIYer. Others are great ideas that would be difficult to retrofit into an existing structure, but just might be a cinch for the enterprising coop-builder to integrate into a basic design for one totally tricked-out coop.

Electricity

So many chicken-folk talk wistfully about how awesome it would be to have electrical power running to their coop that we dedicated a whole chapter to the idea (see Chapter 11). It makes a lot of sense. Having plentiful light at the flick of a switch sure makes late-night or early-morning chicken chores easier to tackle. And an outlet or two gives you the opportunity to plug in a small heater when the weather turns bitter and frigid, or a turbine fan when the air inside the shelter gets a little stinky and stale.

Need more reasons why running power to the coop might be worthwhile?

>> You might appreciate being able to plug in a radio (or your MP3 player speaker dock, if you're a next-gen chicken-raiser) to make your next coop cleanout day a little more enjoyable.

>> Maybe you'd benefit from having juice to run your power tools for some needed repair work or a coop add-on project.

>> You might consider installing an electric fence around the top of your run to combat climbing predators. (Hey, more than one chicken owner has done it.)

>> Maybe you'd just like to tuck a simple mini-fridge in a corner — you know, to keep certain chicken medications (or a small stash of your favorite adult beverage) cold.

Running power to your coop doesn't seem so shockingly overdone *now*, does it?

Solar Power

Your coop just sits out there in the yard all day long, soaking up the sun. Why not put those rays to good use? Alternative energy like solar power is all the rage, and not just for humans' houses. Even a small solar panel mounted on the roof of a coop can be used to save and store enough solar energy to power lights inside the shelter or keep water in a watering dish from freezing during cold winter nights. Solar power isn't necessarily inexpensive to start harnessing, but it may help you pinch enough pennies in the long run to be worthwhile. (And starting small with solar — like on a chicken coop — may help you decide whether it's right for your own home. Who says we can't learn anything from our chickens?)

Running Water

Plumbing your chicken coop is one of those little luxuries that doesn't seem so wildly decadent once you stop to think about it. For starters, it would make cleaning the coop a lot easier (not to mention cleaning *yourself* up before you go back into the house). A nearby tap would also be great for making sure that your birds always have fresh, clean, drinking water.

TIP

While actually calling a plumber to install water supply lines probably qualifies as overkill, many coop-owners have found some clever ways to make running water at their coop more than just a pipe dream. In Chapter 2, we show you a faucet extender that puts a garden hose spigot right where you want it. Other coop-owners have rigged up rain barrel systems (check out Figure 18-1) that catch rainwater as it sheets off the coop roof (via gutters and downspouts) and collect it in large drums. This water can then be used as needed by turning a valve at the bottom of the drum. (It's low-pressure water, but still better than hauling pailfuls back and forth to and from the house!)

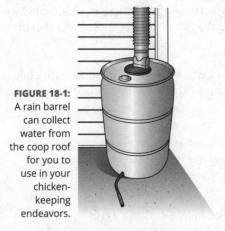

FIGURE 18-1: A rain barrel can collect water from the coop roof for you to use in your chicken-keeping endeavors.

Storage Space

As mentioned in Chapter 17, most coop-owners end up wishing they had a bigger coop. But not all of them crave that extra floor space for more chickens. The longer you keep chickens, the more stuff you accumulate. Supplies, feed, medicines, various gadgets and gizmos: They all need a place to call home. And many caretakers find themselves carving out room in the garage, the basement, the toolshed, or other areas of the house to stash it all. Wouldn't it be great if it all lived where the chickens do? After all, that's where you'll use it.

As you design your coop, look for nooks and crannies where you can work in a storage bin here or an old wall cabinet there. If space allows, you might even consider a dedicated storage room in a walk-in coop. It'll make your job as a chicken-keeper a lot easier.

A Quarantine Area

The problem with constructing a large coop is that it's all for the birds (as far as the birds are concerned). No matter what your plans may be for part of that shelter space, eventually, they'll just take over the whole thing — if you let them.

Why not create a dedicated portion of the coop for birds who may need to be separated from the flock? It happens to every caretaker from time to time: One hen comes down with something and threatens to pass it along to the rest of her housemates; one bird is having some anger management issues and needs a time-out; you get the picture. As long as we're talking about the ultimate coop here, you'd be hard-pressed to find an owner who wouldn't agree that a dedicated "quarantine" area would be a mighty nice thing to have. All it takes is a few extra square feet of floorspace inside the shelter; why not just include it in your design from Day One?

An Automatic Feeder and Waterer

It's funny how as soon as you bite the bullet and build a coop for your new chicken hobby, you immediately start looking for ways to spend less time trudging out there to tackle chicken-related chores. Going out every day to feed and water the flock is one of those tasks that often loses its luster after a while.

Sure, you can buy automatic feeders and waterers, but why? If you can build an entire housing structure for your birds, then cobbling together one of these should be child's play. A few popular designs can be seen in Figure 18-2.

>> **Automatic feeders:** A large plastic bucket with holes cut out around the bottom (see Figure 18-2a) or a wooden bin that empties into a tray (see Figure 18-2b) makes a great self-feeder for your chickens and helps keep the feed contained, minimizing waste and mess inside the coop.

>> **Automatic waterers:** Automatic watering systems often feature nipple valves that drip water when a bird touches one with its beak. You can buy these nipples and fashion your own plastic bottle, bucket, or jug into a perfectly good watering station (see Figure 18-2c). Other waterers simply hold a large volume of water in a tank and keep a drinking tray filled to a certain level (see Figure 18-2d).

a

b

c

d

FIGURE 18-2: Automatic feeders and watering systems save time.

WARNING

Don't get too complacent when using an automatic system! These systems still need to be checked on a regular basis to make sure they're filled with enough feed or water, that they're clean and accessible to the birds, and that they're in good working order at all times.

TIP

Here's one clever idea we heard about: A chicken owner wanted to add a little light and some warmth to his coop, which was outfitted with electrical power. He screwed a 60-watt bug bulb into the coop's light socket, and voilá! His chicks also had their own insect buffet within minutes. The yellow bulb draws bugs to it and acts as its own kind of "automatic feeding system." And the caretaker gets a free show every night, watching his flock jump and peck at the swarming pests.

A Removable Droppings Pan

Any wish list for tricking out the ultimate coop has to include at least one item that makes cleaning chicken waste at least a little easier. Try building a removable droppings pan. This tray sits underneath the roost bar and collects droppings on an easy-to-clean surface like linoleum or Glasbord. A hatch in an exterior wall allows the entire thing to be removed from outside the coop, where it can be hosed or scraped off, and then slid right back into place. You'll thank us every time you use it. (Check out Chapter 9 to see one.)

Wheels

If you're building a large walk-in coop or a stationary shelter that sits on posts anchored in the earth, you can skip this wish-list item. But builders of most other small coops may want to consider putting their shelter on wheels. Sure, a true tractor coop is meant to be moved around and usually incorporates wheels or a set of skids (see Chapter 14 for an example of a tractor coop). But even if you don't intend to relocate your henhouse on a regular basis, you can probably envision times when it would be convenient to move it out of the way.

Maybe you're having a backyard barbecue or you just want to mow the lawn. Or perhaps a forecasted cold snap has you thinking about relocating your chickens to the garage for a few nights. A coop that's portable sure gives you more options than one that you can't budge. You may move it only once or twice, but that just might make it worth checking eBay for a couple of heavy-duty, all-terrain tractor tires. Slap 'em on at a moment's notice, and wheel your coop away should the need arise. (Sure sounds handy to us. But that's just how we roll.)

A Retractable Roof

If they can build football stadiums with retractable roofs, then putting one on a chicken run shouldn't be too hard. Think about it: With a little 2x4 framing, some corrugated roofing panels, and a few heavy-duty hinges, it's not difficult to imagine a chicken run that can go topless on bright sunny days, and then be closed up at night or when rain moves in. And although we've never actually seen a run with a retractable roof, we're intrigued enough by the idea to put it on our ultimate wish list.

"Air-Lock" Doors

Sooner or later, it happens: You head out to the coop one morning, open the door, and one lone bird who may or may not have been waiting at that door all night long slips out and makes a break for it. Suddenly, you're doing your own private reenactment of the chicken-chase training scene in *Rocky II*. All that's missing is Burgess Meredith calling you a bum and screaming something about lightning and thunder.

TIP

It's right about then that this idea will seem like pure, solid gold. A few clever coop owners have installed a second door to their coop or run, just outside the first one. This creates an antechamber or "air-lock" effect — you leisurely go in the first door while the chickens are still safely confined behind the second. Only after you close the first door behind you do you open the second one. If you have a runner, she'll only get as far as the first door. Cool, huh? Now head to the nearest set of City Hall steps for your best Rockyesque sprint to the top, arms pumping overhead to celebrate your genius for implementing this idea.

Automatic Door Closers

It takes a certain kind of tech nerd to pull this one off. Or a chicken-owner who would prefer to sleep in rather than traipse outside just to open the coop door and allow his birds to head into the run to get an early start on a long day of being chickens.

A surprising number of coop-builders have figured out ways to get that door to open in the morning and close at night all by itself, using transistors, relay switches, and various electronic parts and pieces from Radio Shack. Some of these automatic doors work on timers, while others use photoelectric eyes that sense when the sun has come up or gone down. But the idea is that the door automatically opens early in the morning to let the chickens out, and then closes at night after all the chickens are known to be safely inside the coop.

This might be a great addition to your coop if you have a penchant for small electronics, soldering, and wiring projects. Or if you know a whip-smart 8th-grader who'd be willing to build it for you and then let you keep it after he rocks this year's science fair.

A Wireless Weather Station

Sure, you can find out how cold or hot it is outside with a few clicks of your computer mouse, a quick scroll with your TV remote, or by stepping out onto your front porch. But do you really know what the temperature is inside your chicken coop? Sun, shade, wind, and even the chickens' body warmth can all make a noticeable difference; it may be quite a bit colder or warmer in there than you think.

With a wireless weather station, you place a small unit inside the coop. It then transmits the temperature (and other info like humidity) to a display that you keep conveniently located inside your house. You'll know at a glance if your chickens are in danger of getting too hot or too cold inside their shelter. Wireless weather stations are relatively easy to find in many department stores, shops that specialize in small electronics, those quirky gift catalogs you get in the mail or on an airplane, and from many online retailers.

Index

hawk, 20, 71
H-clip, 141, 142
head, screw, 66
header, 118, 119–120
heater, 176
helix nail, 62
hinge
  door installation, 137–138
  nest boxes, 156
  roosts, 153
  window options, 139
hoe, 44
homeowners' association, 24
hoop coop, 29–30
hose, 26, 27
hot-dipped nail, 62
hot-dipped screw, 65
house, 24–25
hurricane clip, 128

I

icons, explained, 5
insect attractor, 273
insulation, 176

J

jack stud, 118
J-channel trim, 135
jig, 90
jigsaw, 40, 133
joinery, 89–90
joist
  purpose of, 14
  roof installation, 125, 126
  subfloor installation, 110–113
junction box, 171, 172

K

kerf, 81
kickback, 80
king stud, 118, 119
knife blade, 51
knot, in boards, 56
Kreg Tool Company, 90

L

ladder, 127
lag screw, 65, 66
laminated wood veneer, 58
landscape
  coop location, 24, 25, 37
  site prep, 37, 98
latch, 138, 139
law, zoning, 24
length measurement, 38
level
  basic equipment, 11, 48–49
  carpentry skills, 13
  defined, 13
  leveling tips, 13, 48–49
  post installation, 43–44, 102, 107
  roof installation, 126, 141
  slope of ground, 98–99
  types of, 48–49, 92–93
  wall installation, 134
lighting
  artificial light, 173–174
  coop basics, 10
  coop location, 26–27
  fixtures, 174–175
  insect attractors, 273
  light placement, 174
  nest box requirements, 22, 154–155
  roofing materials, 70, 145
  timers, 175
  types of, 20
line level, 48, 92, 99
linear feet, 73
linoleum, 67, 149
Liquid Nails (adhesive), 113
litter box, 152, 157
location, of coop
  selection of, 23–28, 265
  site prep, 37, 97
lock
  door installation, 138
  tape measures, 38
long point, 193
lubricated nail, 87

Ludlow, Rob (*Raising Chickens For Dummies*), 3
lumber. *See also specific types*
  All-in-One coop, 223–237
  Alpine A-Frame, 192–202
  board sizing, 54–55
  carpentry skills, 13
  defined, 54
  estimates quantity, 73
  grading system, 55
  measuring tools, 37–39
  Minimal Coop, 182–187
  recycled materials, 53–54
  roost materials, 152–153
  selection, 56–60
  sheet sizing, 58
  types of, 55–56, 58–60
  Urban Tractor coop, 206–216
  Walk-In coop, 244–255

M

marking items. *See also specific tools*
  carpenter's skills, 13, 76–78
  tools, 38–39, 92
masking tape, 85
materials, basic. *See also specific items*
  basic requirements, 11–12, 53
  costs, 265
  quantity estimations, 55, 73
math skills, 13, 76
mattock, 37, 98
MDF (medium–density fiberboard), 60
measuring items. *See also specific tools*
  carpenter's skills, 75–76
  tools, 37–39, 49
medium-density fiberboard (MDF), 60
mending plate, 90
metal bracket
  floor installation, 111, 112
  post installation, 105–106
metal post, 72, 163
mildew, 177
milled face, 45

Minimal Coop
  construction process, 187–189
  described, 17, 181–182
  materials list, 182–187
miter saw
  cutting process, 82–83
  described, 39–40
mounted post, 105–106

N

nailing
  decking installation, 113
  described, 60–61
  door/window installation, 120, 137
  estimating quantity, 73
  hammering hints, 85–86, 88
  hammer's face, 45–46
  illustrated, 63
  joinery tips, 88–89
  nail guns, 46
  pulling tips, 86
  roof installation, 141, 143, 144
  roosts, 153
  run installation, 165–169
  versus screws, 61, 64, 266
  sizes, 61, 64
  subfloor installation, 111–113
  types of, 62–63
  wall installation, 115, 117, 120–121, 135
  wood splitting, 87
nail gun
  basic equipment, 11
  described, 46
  versus hammer, 46
nailing cleat, 153
neighborhood covenant, 24
neighbor's house, 24–25
nest box
  accessibility, 267–268
  All-in-One coop, 232–233, 240
  Alpine A-Frame coop, 192, 198–199, 203
  construction process, 156–157

described, 15, 22
design, 154–156
general rules of, 22
illustrated, 22
Minimal Coop, 182, 186–187, 188
purchased boxes, 157
roost location, 151
Urban Tractor coop, 206, 214–215, 218–219
Walk-In coop, 251–254, 259
noise, 25
nominal dimension, 54
notching posts, 111
nylon fencing, 71

O

Odjob (tool), 104
odor, 25
one-by-four (1x4) board, 55
one-by-three (1x3) board, 55
one-by-two (1x2) board, 54, 55
opossum, 20
orange snow fencing, 71
organic chickening, 57
oriented strand board (OSB)
  described, 59
  wall installation, 134, 135
outlet, 26, 171–173
overhang, 127
owl, 20, 21, 71

P

paint, 268
palm nailer, 47
pan, dropping
  cleanliness of coop, 67
  roost installation, 152, 273
panel
  doors, 136
  fencing, 72
  roofs, 141, 145
  runs, 165
  walls, 116–117, 130–136
  windows, 139–140

particle board, 59–60
pencil
  basic equipment, 11
  described, 38
  marking skills, 13
penny weight, 61, 62
pent roof, 123
Phillips-head screw, 66
pier block, 105–106
pilot hole
  door/wall installation, 134
  fastener types, 65
  joinery tips, 89–90
pitch, 123–124, 126
pivot point, 94
plants, 98
plastic fencing, 70–71
plate
  door/window installation, 119
  joinery tips, 90–91
  roof installation, 125–126, 128
  wall installation, 114, 116–117, 120, 121
plumb board
  carpenter's skills, 13
  post installation, 102–103
  tools, 48, 91–92
  wall installation, 122
plumb cut, 127
plumbing
  benefits of, 270–271
  coop location, 25–26
plywood
  blades, 79
  cutting process, 84–85
  cutting skills, 79
  decking installation, 113
  described, 58–59
  nest boxes, 156
  roofing materials, 68–69, 141
  wall installation, 130–133
  wall materials, 68
pneumatic nailer, 46
pocket hole, 89–90
pollution, 27

About the Authors

Todd Brock: Todd Brock has written, directed, and produced more than 1,000 episodes of television programming. His shows on topics ranging from landscaping to home renovations to gardening have been broadcast nationally on major networks including HGTV, DIY Network, and PBS, and locally in one of the country's Top 10 TV markets.

As a freelance writer, Todd has researched and written about everything from mobsters to Pac-Man, and children's stories to cheeseburgers. He lives in the Atlanta, Georgia, area with his wife, Debbie, and their two daughters, Sydney and Kendall.

Dave Zook: Dave Zook, his wife, Suz, and their four children, Justin, Jordan, Jenika, and Javon, live on several acres in rural Lancaster County, Pennsylvania. He is the founder/owner of Horizon Structures, a manufacturer of pre-built storage sheds, garages, horse barns, and chicken coops.

Dave and his family keep a small flock of chickens at home in one of his company's coops. He continues to improve the designs and develop new ones based on customer input as well as his family's experiences with their own backyard flock.

Over the past nine years, Horizon's line of chicken coops has proven to be very popular with chicken fanciers — and their hens — throughout the U.S., with coops now in 48 states!

Rob Ludlow: Rob Ludlow, his wife, Emily, and their two beautiful daughters, Alana and April, are the perfect example of the suburban family with a small flock of backyard chickens. Like countless others, what started out as a fun hobby raising a few egg-laying machines has almost turned into an addiction.

Originally, Rob started posting his chicken experiences on his hobby Web site, `www.Nifty-Stuff.com`, but after realizing how much his obsession was growing, he decided to concentrate his efforts into a site devoted completely to the subject. Now Rob owns and manages `www.backyardchickens.com`, the largest and fastest-growing community of chicken enthusiasts in the world.

Rob is also the coauthor of the book *Raising Chickens For Dummies.*

Dedication

Todd Brock: I dedicate this book to Debbie, Sydney, and Kendall — the three crazy chickens living in my coop.

Dave Zook: Thanks to my wonderful wife, Suz, and to my kids, Justin, Jordan, Jenika, and Javon, for your patience, sacrifice, and help as I studied and got the details together for this book.

Rob Ludlow: To Mom, Dad, and my five older siblings for supporting me in the very diverse areas of my life that range from Web site design to raising backyard chickens.

Authors' Acknowledgments

Todd Brock: Thanks to my project editor, Georgette Beatty, for guiding me through my first *For Dummies* book, to copy editor Christy Pingleton for cleaning up my messes, and to acquisitions editor Mike Baker for entrusting the project to me in the first place. Special recognition goes to technical reviewer Terry Schmitt and to the composition staff for working tirelessly to proofread and lay out this book. Precision Graphics did an incredible job in creating the art for the coop plans, and Wiley's graphics department is credited with producing the rest of the superb art in these chapters. To my coauthors Rob and Dave, thanks for graciously answering all my dumb questions along the way. Finally, my sincerest gratitude goes to Lindsay Lefevere for bringing me into the *Dummies* fold to begin with.

Dave Zook: I want to thank my team at Horizon Structures — especially Matt and Dan — for all the time-consuming, and often tedious, work involved in gathering all the details, sketching the coops, and making sure everything was perfect for this book. Thanks to Jill for your good ideas on coop designs and your work on getting the information about our coops out to the public. Thanks also to Todd Brock, Mike Baker and Georgette Beatty from Wiley, and Rob Ludlow from BackYardChickens.com for everything you did to make all of this possible.

Rob Ludlow: Thanks to my church leaders, family, friends, and Boy Scout leaders for teaching me how to build "stuff." Thanks also to Mike Baker for putting up with my endless questions and suggestions, and to the thousands of BackYardChickens.com community members who continue to show me how amazingly simple or complex, plain or ornate, and cheap or expensive a chicken coop can be!

Publisher's Acknowledgments

Senior Project Editor: Georgette Beatty

Senior Acquisitions Editor: Mike Baker

Copy Editor: Christine Pingleton

Assistant Editor: Erin Calligan Mooney

Senior Editorial Assistant: David Lutton

Technical Editor: Terry Schmitt

Editorial Manager: Michelle Hacker

Editorial Assistant: Jennette ElNaggar

Art Coordinator: Alicia B. South

Cartoons: Rich Tennant (www.the5thwave.com)

Production Editor: G. Vasanth Koilraj

Cover Image: © richardernestyap/Shutterstock

Take dummies with you everywhere you go!

Whether you are excited about e-books, want more from the web, must have your mobile apps, or are swept up in social media, dummies makes everything easier.

Find us online!

dummies.com

dummies®
A Wiley Brand

PERSONAL ENRICHMENT

9781119187790	9781119179030	9781119293354	9781119293347	9781119310068	9781119235606
USA $26.00	USA $21.99	USA $24.99	USA $22.99	USA $22.99	USA $24.99
CAN $31.99	CAN $25.99	CAN $29.99	CAN $27.99	CAN $27.99	CAN $29.99
UK £19.99	UK £16.99	UK £17.99	UK £16.99	UK £16.99	UK £17.99

9781119251163	9781119235491	9781119279952	9781119283133	9781119287117	9781119130246
USA $24.99	USA $26.99	USA $24.99	USA $24.99	USA $24.99	USA $22.99
CAN $29.99	CAN $31.99	CAN $29.99	CAN $29.99	CAN $29.99	CAN $27.99
UK £17.99	UK £19.99	UK £17.99	UK £17.99	UK £16.99	UK £16.99

PROFESSIONAL DEVELOPMENT

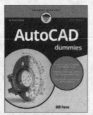

9781119311041	9781119255796	9781119293439	9781119281467	9781119280651	9781119251132	9781119310563
USA $24.99	USA $39.99	USA $26.99	USA $26.99	USA $29.99	USA $24.99	USA $34.00
CAN $29.99	CAN $47.99	CAN $31.99	CAN $31.99	CAN $35.99	CAN $29.99	CAN $41.99
UK £17.99	UK £27.99	UK £19.99	UK £19.99	UK £21.99	UK £17.99	UK £24.99

9781119181705	9781119263593	9781119257769	9781119293477	9781119265313	9781119239314	9781119293323
USA $29.99	USA $26.99	USA $29.99	USA $26.99	USA $24.99	USA $29.99	USA $29.99
CAN $35.99	CAN $31.99	CAN $35.99	CAN $31.99	CAN $29.99	CAN $35.99	CAN $35.99
UK £21.99	UK £19.99	UK £21.99	UK £19.99	UK £17.99	UK £21.99	UK £21.99